BOSTON'S FIRE TRAIL

Assistant Chief John W. Regan typifies the image of a chief fire officer for that time period. Born in 1835, he was promoted to assistant chief on May 1, 1887, and retired on October 29, 1897. Note the three crossed trumpets signifying his rank.

A WALK THROUGH THE CITY'S FIRE AND FIREFIGHTING HISTORY

BOSTON'S FIRE TRAIL

BOSTON FIRE HISTORICAL SOCIETY, WITH A FOREWORD BY PAUL A. CHRISTIAN

Charleston   London
History
PRESS

Published by The History Press
Charleston, SC 29403
www.historypress.net

Cover image: The February 1, 1976 fire at the Plant Shoe Factory in Jamaica Plain.

First published 2007

Manufactured in the United Kingdom

ISBN 978.1.59629.361.8

Library of Congress CIP data applied for.

Notice: The information in this book is true and complete to the best of our knowledge. It is offered without guarantee on the part of the authors or The History Press. The authors and The History Press disclaim all liability in connection with the use of this book.

Contents

Contents

Foreword

BOSTON'S FIRE HISTORY TRAIL

As one of the nation's oldest big cities, Boston has been the birthplace of many progressive historic, social and political movements. Many of the country's great institutions can be traced to traditions established by colonists who built a community on a hill overlooking a deep-water harbor. One of those traditions concerned public safety: it was a Boston citizen's duty to look out for his fellow citizens; helping those in need would benefit everyone. Thus, it is no coincidence that the Boston Fire Department enjoys a reputation of being the oldest and one of the finest fire departments in the country.

Serious fires were frequent in Boston's colonial period. Early construction in the town of Boston was dense and combustible, and fire was the most feared enemy. Conflagrations, with devastating results, led to the establishment of the first fire companies and first fire prevention regulations in the country. Fire buckets and other firefighting implements were mandated to cope with the hazard. Fire companies were formed that, in addition to providing a valued public service, also proved to be powerful social and political organizations.

From its organization in 1678, the Boston Fire Department, through ingenuity and necessity, grew into a powerful firefighting force noted for its dedication, innovation and bravery. It has long served as a model for municipal fire departments and its members have been sought to lead other fire departments over the years. Many of the department's traditions can be traced back to the first fire volunteers who battled blazes while Boston was still a colony.

In the early nineteenth century, bucket brigades and hand-drawn "tubs" gave way to more efficient hand-drawn suction engines. Just before the Civil War, horse-drawn apparatus and steam fire engines came into use. By the 1920s, horses were replaced by motor fire apparatus, which became a diesel-powered fleet late in the twentieth century.

In its 325-year history, the city of Boston has been the scene of countless significant fires and disasters. The daring and bravado of Boston's firefighters is legendary. Recognized for innovative firefighting techniques and for having the nation's most skilled

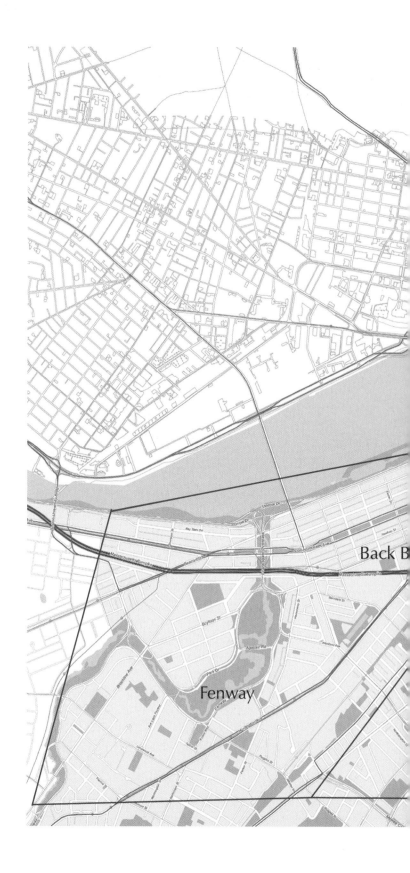

Charlestown

North End

Downtown/
Financial District

South End

Boston's Neighborhoods

laddermen, Boston was the site of many "firsts" in the American fire service: the first fire engine, the first paid fire department, the first electric telegraph fire alarm system and the first use of radio in the fire service are among them.

Perhaps the greatest aspect of the Boston Fire Department has been the courage and dedication of those who have served in the BFD uniform. The record of the department is filled with accounts of heroism and sacrifice, often made at great cost in human terms, to carry out its mission of service and safety.

While the city of Boston has undergone many changes through urban renewal and expansion, much of "Old Boston" remains. Though they are now wider, many of the original streets and alleys are still in place. Most of the sites of fires and disasters are readily accessible and identifiable. You can stand near the location of landmark events such as the Great Boston Fire or the infamous Cocoanut Grove Fire or at the Vendome Firefighters Memorial and imagine the trauma, tragedy and triumphs that once occurred before you. It is the goal of this book to recall and preserve some of the significant events and fires that shaped the Boston Fire Department—information that, otherwise, would be lost through the passage of time.

In these pages, we will walk a "fire trail" that traces the history of the Boston Fire Department. You can use this book as a guide when visiting Boston, which is one of the nation's most "walkable cities." Or you can relax in an armchair and follow the stories as they tell the history of the Boston Fire Department. As you read, imagine that you can hear the shouts and clatter of the earliest firefighters running through the small twisting streets with their "masheens" and leather buckets; listen as "call" men race with their hand-drawn apparatus with clanging bells. You might hear the beat of horses' hooves on the old cobblestone streets as well-trained steeds speed to the alarm, or the cadence of the public bells striking out an alarm. Imagine, while looking over a modern busy street or from the safety of your armchair, the choking smoke and searing heat of the flames; the wind and ice of bitter Boston winters; hear the sounds of a fire fight, the breaking glass, the rush of powerful water streams and the thunderous crash of collapsing walls.

It is all here on Boston's fire trail. Let's take a walk through history.

Paul A. Christian
Fire Commissioner, Retired

How to Use This Book

Boston is a confusing city—even to those of us who have lived here for years—so to help guide you on Boston's "fire trail," we have created four chapters, each focusing on a different neighborhood, and a fifth chapter with sites and fires outside of central Boston. When visiting the first four neighborhoods, we suggest leaving the car parked and walking; all the sites within a neighborhood are easily accessible by foot.

The entries are listed roughly in order of geographic location; please consult the map if you wish to wander from site to site. We suggest that you take your time and meander as much as you want.

This book is meant to be a combination history book and guidebook; you can even use it alongside maps or guides to the city's famed Freedom Trail. Members of the Boston Fire Historical Society put their heads together to come up with the fires and other sites listed here. Each member brought a different expertise to this project, so the entries vary accordingly and are each tagged with the initials of the author.

One final note: space and time make it impossible for us to include all major fires, firehouses or other locations of significance to the Boston Fire Department. Our goal is to give readers—whether the public or firefighters—a taste of the amazing history of firefighting in Boston.

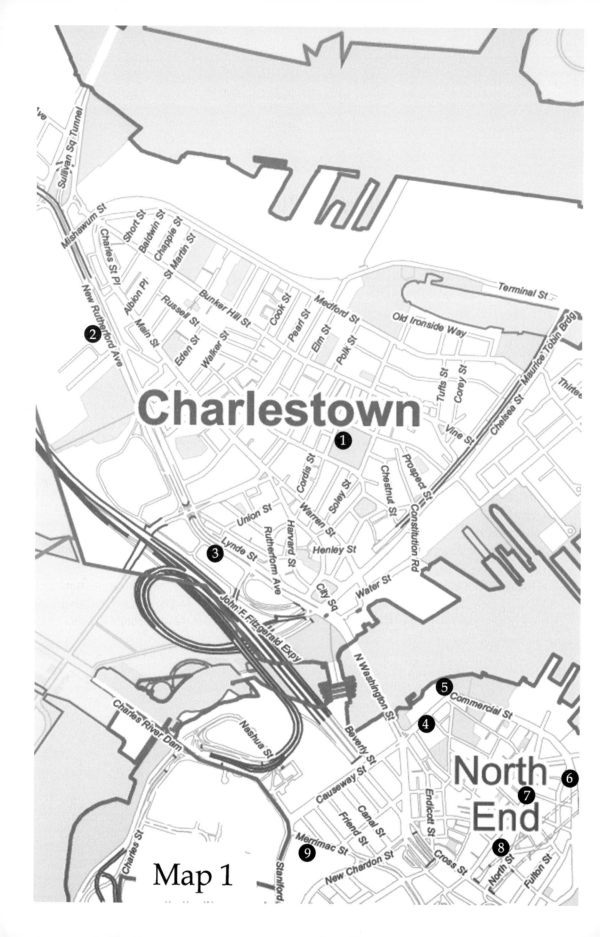

Map 1.

Charlestown/North End

1. The Burning of Charlestown

Bunker Hill Monument, Charlestown
The red line of Boston's Freedom Trail winds past sites important to American history. But that red line marks sites that are important in Boston's fire history as well. One of the most familiar of all American monuments is the distinctive Bunker Hill obelisk, which commemorates the Battle of Bunker Hill. Fire played a role in that landmark event. The battle was the first major armed conflict for the Patriots in the American Revolution; it took place on Breed's Hill on June 17, 1775. The battle was launched when Patriots began to build fortifications on the hills of Charlestown, which was then a town outside Boston. British General William Howe began to move in for an attack, and about 2:00 p.m. he ordered his naval forces to fire heated shot into Charlestown. Nearly four hundred buildings caught fire and burned for hours, devastating the town. The wholesale destruction, which was witnessed by many watching the battle from hills and roofs in Boston, increased antagonism against the British and helped to bring on full-scale revolution. Patriot forces actually lost the Battle of Bunker Hill, but they inflicted such high casualties on the British that all parties realized that the fledgling Revolutionaries were more than a ragtag group of troublemakers, but rather the forefront of a growing movement.
[S.S.]

2. Rutherford Avenue Conflagration

Rutherford Avenue, near Middlesex Street, Charlestown
Thursday, September 18, 1941, was a day that would long be remembered for any firefighter working and anyone living in Charlestown. This was the day of the Rutherford Avenue Conflagration. Just before 6:00 p.m., the Fire Alarm Office received a telephone call from the Columbia Radiator Company at 450 Rutherford Avenue for a

A fire on the Warren Avenue Bridge in Charlestown on September 24, 1935, caused major damage. The Charles River Dam occupies the area where the bridge was located. Note the Bunker Hill Monument at the top right of the photo.

A spectacular fire occurred on September 18, 1941, in the railroad freight sheds off Rutherford Avenue in Charlestown. Note the firefighter at the firebox using the fire alarm telegraph to send for more help. Two-way fire service radios did not come into general use until after World War II.

fire in the area; Engine Company 32, Ladder Company 9 and District 2 were dispatched at 5:59 p.m. Box 4156 at Rutherford Avenue and Middlesex Street was transmitted at 6:02 p.m., bringing Engine 50, Engine 27 and Ladder 22, the balance of the first alarm assignment.

Engine 32 proceeded to the rear of the freight shed at 470 Rutherford Avenue, where a lively fire was burning. The fire originated in or around a trailer truck parked at the Boston & Maine Railroad Freight House Number 35, which was occupied by First National Stores as a warehouse for canned food. The fire spread to several freight cars; it had a good hold on the shed and was extending rapidly toward the front of the building at Rutherford Avenue. The battle was on as the fire moved with great speed, and in no time the entire area was fully involved. Fire department forces were set up to stop the fire from spreading in a southerly direction.

Freight Shed No. 35—a one-story wood structure 620 feet long and 60 feet wide, with 18-foot-high ceilings—was one of many similar structures that lined the west side of Rutherford Avenue for about a half-mile between Sullivan Square and the Massachusetts State Prison at Prison Point Bridge. As the fire was spreading rapidly and covered an estimated twelve acres, District Two Chief Patrick E. Collins skipped the second alarm and ordered the third alarm by telegraph from Box 4156 at 6:06 p.m. The fourth alarm was ordered at 6:07 p.m. and the fifth alarm at 6:12 p.m. A call for additional engine companies was ordered at 6:26 p.m.

The fire consumed everything in its path, engulfing numerous freight sheds along Rutherford Avenue. These buildings were all of similar size and construction as Shed 35, averaging over six hundred feet in length, and were used for storage of roofing materials, hardwood flooring, radiators, enamelware pottery and cases of canned food.

Fire companies had trouble getting into position at the fire. Numerous freight tracks crossed the area between the freight sheds, as well as at the rear of the complex occupied by Boston & Maine Railroad freight yard No. 19. Engine companies had difficulty getting access to hydrants at the western end of the freight houses due to the tracks, the terrific heat and the lack of roadways. Engines eventually connected to hydrants and got water on the fire in the rear. Apparatus and men had to be repeatedly repositioned as the fire bore down on them and sailed over their heads.

After consuming about twelve acres, the main body of fire was eventually stopped with heavy streams; a 150-foot-wide driveway acted as a firebreak.

On the orders of Acting Chief of Department Louis C.I. Stickel, signal 10-21 was sounded at 6:53 p.m., which was the "recall" signal for all off-duty members to return to duty. Radio stations were also notified to ask off-duty firefighters to return to work.

The all out was sent at 12:47 p.m. on September 19, eighteen hours and forty-five minutes after the first alarm was sounded. The department had used 34,600 feet of hose at the fire and lost over 7,000 feet, burned in the fast moving fire. The fire destroyed twenty-one railroad freight cars and eight tractor-trailer units, plus numerous freight shanties.

The conflagration was the largest in the city since the Blacker and Shepard fire of August 1910. A total of fifty-six engine companies (thirty-six from Boston and twenty via Mutual Aid), eight ladder companies, two rescue companies and two water towers responded. The total damage was estimated at $1,576,000.

Over the years, the area of the fire has been transformed. Taking the place of railroad yards, freight houses and industrial buildings are new roads and the sprawling campus of Bunker Hill Community College. Rutherford Avenue has also been realigned.
[T.G., J.T.]

3. Potato Sheds Fire

Rutherford Avenue, near City Square, Charlestown
On the afternoon of May 10, 1962, one of the city's most spectacular and costliest fires raged for several hours. The fire devoured freight sheds, railroad cars, tenements, a large four-story mill building and anything else that was in its path. The fire originated in the Boston & Maine Railroad freight yards in an area known as Potato Row, located off Rutherford Avenue near City Square in Charlestown.

Radiant heat ignited these dwellings opposite a raging fire in the potato sheds off Rutherford Avenue in Charlestown. This spectacular 1962 five-alarm fire started in a 1,378-foot-long shed. Firefighting was hampered by railroad tracks and freight cars.

The fire started under one of two large sheds, number 18—a one-story, 30-foot by 1,378-foot metal-clad structure—and quickly spread. When the Fire Alarm Office was alerted of a spreading blaze by a member of the Boston & Maine Railroad security, Engine Company 50, Ladder Company 22 and District Chief Robert Beltramini of District 2 were dispatched. Beltramini ordered Box 4132 at Union and Washington Streets struck at 4:01 p.m. Within half an hour, five alarms had been ordered. Soon the fire was so hot that its radiation was igniting all combustible material nearby. Next to Shed 18, Shed 17, built of brick, suddenly burst into flames. The heavy rush-hour traffic on the elevated Central Artery caused delays for the multiple-alarm apparatus due to the large plume of smoke rising from the fire. Chief of Department John A. Martin and the other officials responding were caught up in the traffic; the sight of smoke caused drivers to slow down on the busy highway.

Companies were ordered to set up heavy-stream appliances as the fire spread to the northwest toward Prison Point Bridge. They had to cope with a lack of water due to the small size of the water mains and the spacing of hydrants in the freight yard. The fire grew so intense that Division 1 Deputy Fire Chief Frederick Clauss twice had to order retreats.

Chief Martin arrived at the fire just after the fifth alarm had been sounded and established a command post at Lynde and Union Streets, which permitted better overall appraisal of the fire area. The four-story brick flour mill building in the row with Shed 17 caught fire, and the firefighters manning Engine Company 32's hand line had to abandon their position. Suddenly a three-story frame dwelling at 35 Lynde Street to the north of the fire exploded in flames. The roofs and rear portions of the buildings fronting on Washington Street, numbers 40 through 64, also ignited.

"Hold your positions as best you can and operate on the buildings on Lynde Street and Washington Street," arriving companies were told. Firefighters set up "water curtains," which cut down on the radiant heat, and the fire was stopped there.

Engines were set up to draft from the saltwater canal at the southeast end of the freight yard and supplemented the water supply to pumpers located closer to the fire. Other apparatus were directed to enter Arrow Street, which was a narrow extension of Lynde Street to Rutherford Avenue on the easterly end of the fire. These companies took up positions and operated heavy stream appliances, which were supplied by the large water mains along Rutherford Avenue. They were able to halt the spread of the blaze in that direction. Fortunately, despite the huge property damage, no deaths were reported. The smell of smoke and roasting potatoes permeated the neighborhood for days.

The area of the fire is now part of Interstate 93 and Bunker Hill Community College. There is little left that would indicate there were once large railroad sheds in this area.

[B.N.]

4. Brink's Garage

165 Prince Street, near corner of Commercial Street, North End
In the early evening of January 17, 1950, seven masked men entered the North Terminal Garage, which housed the offices and vault of the Brink's Inc. armored car company. In twenty minutes, bandits subdued guards and made off with about $2.7 million in cash, bonds and securities; this constituted what was then the biggest holdup in the nation. The sensational "Crime of the Century" made headlines around the world. Coincidentally, that night there was a fire near the corner of Atlantic and Commercial Streets, just a few blocks away. For years, some wondered if the fire was set as a diversion. It wasn't, but it would be six years before the Brink's bandits were caught and convicted, and then only because one decided to rat out the others. The Brink's company moved from the garage years ago, but on the door at 165 Prince Street, you can see the faint outline of the Brink's badge that once adorned the surface.

[S.S.]

The molasses tank collapse caused severe damage to the adjacent firehouse of fireboat Engine 31 on January 15, 1919. In this photo, firefighters work desperately to free Third Engineer George Layhe, who was trapped in a pool of viscous molasses. Layhe died before rescue efforts could free him.

5. The Molasses Disaster

529 Commercial Street, opposite Copps Hill Terrace, North End
About 12:45 p.m. on January 15, 1919, a fifty-foot-tall molasses tank containing over two million gallons of molasses, weighing over twenty-six million pounds, ruptured. A fifteen-foot wave of molasses poured out into the street, trapping and killing twenty-one persons and injuring forty. The molasses tank had been built in a crowded section of Boston in 1915, and owners had not completed preliminary tests with water to see if the tank could hold the weight of the molasses. The molasses was originally used in the manufacture of munitions for World War I. Molasses was pumped through pipes from ships into the tank and was later pumped into railroad cars for transfer to munitions plants in Cambridge. However, with the war over and Prohibition (the Eighteenth Amendment) set to take effect in January 1920, tank owners sought to maximize the amount of molasses on hand for the production of rum and beat the new law's restrictions on alcohol production.

When the tank ruptured, a flood of molasses traveling thirty-five miles per hour inundated the area. Two alarms were sounded from Box 1234. The viscous syrup

An elevated view shows the collapsed molasses tank on Commercial Street in the North End. The remains of the tank can be seen on the right. Over two million gallons flowed from the tank, causing a fifteen-foot wave to sweep through the streets at thirty-five miles per hour. The collapse killed twenty-one and injured forty persons.

destroyed the fireboat station and the city paving yard and damaged the uprights of the Boston Elevated Railway and the Union Freight Railroad track and property. Houses and businesses adjacent to the tank were flooded with molasses. An elevated train en route to North Station narrowly avoided falling off the sagging trestle. George Layhe, a third engineer with fireboat Engine 31, and other firefighters were pinned by fallen beams in the collapsed firehouse, which was pushed off its foundation and nearly thrust into the harbor. All were eventually rescued, except Layhe. He kept his head out of the pooled molasses for a time, but he drowned before rescue efforts could free him, according to Stephen Puleo, author of *Dark Tide: The Great Boston Molasses Flood of 1919*. The response by the fire department and other departments was hampered by the sticky molasses, which coated personnel and equipment. Lawsuits against the owners of the tank, United States Industrial Alcohol Company, took nearly six years to settle. The total settlement was for $300,000, about $30 million in current dollars.

Today, the area is the site of the Langone Athletic Complex and Andrew P. Puopolo Jr. Athletic Field, which includes playing fields, bocce courts, a children's playground and a waterfront park. A small marker mounted on a wall beside the sidewalk on Commercial Street is the only sign of the syrupy tidal wave that proved so deadly.
[M.G.]

6. Union Wharf Fire

Off Commercial Street, opposite Murphy Court, North End

Little evidence remains today of Boston's once-thriving commercial waterfront. Old wharves and warehouses have given way to more genteel utilization: cruise terminals, restaurants, condominiums, hotels, parks and marinas. Gone are the wharves, warehouses, shipping line terminals and wooden freight sheds. Many fell victim to major fires during the 1950s and 1960s.

One such wharf located on Commercial Street in Boston's North End was Union Wharf, scene of one of Boston's most spectacular fires on Sunday afternoon, November 2, 1952.

The wharf contained a two-story, two-hundred- by six-hundred-foot, J-shaped, tin-clad wooden freight shed fronting on the Commercial Street side; the shed became a one-story section at the edge of the wharf. Union Wharf was, as the saying goes, "built to burn," and was long feared by Boston firefighters. The wharf was not equipped with sprinklers, firewalls or fire stops under the pier, nor did it have an automatic alarm system. Conditions and fire loading could not have been worse: there were wooden boxes, pallets and hundreds of eight-hundred-pound rubber bales.

October of 1952 had had only an infinitesimal rainfall. Governor Paul A. Dever had ordered the state's woodlands closed for fear of a major forest fire. On the afternoon of November 2, the humidity was very low and a light wind was blowing in from the southwest. The conditions were ideal for a fire.

That Sunday had been very busy for Boston firefighters; there had been a three-alarm fire in Mission Hill shortly after 11:00 a.m. and numerous struck boxes and brush fire activity in the Hyde Park and West Roxbury districts. At 12:42 p.m., several events took place: Box 1242 was received and transmitted; the Boston Police boat *McShane* called the turret at Boston Police Headquarters to report a fire on Union Wharf; the pilot aboard the East Boston ferry *Donahue* spotted the fire and sounded the marine signal; and members of Engine Company 47, a fireboat docked at nearby Battery Wharf, pulled out from its berth and radioed Boston Fire Alarm Headquarters that they could see a fire on Union Wharf and it looked serious.

Deputy Chief Edward J. Gaughan, Division One, arrived from Engine Company 8's quarters on Hanover Street to find heavy fire wrapping around the harbor end of the wharf shed and extending halfway forward on the six-hundred-foot shed. He ordered a second alarm before he got out of his car at 12:44 p.m., followed by a third alarm at 12:46 p.m. The fourth alarm was struck at 12:47 p.m., and Chief of Department John V. Stapleton ordered the fifth alarm on his arrival at 12:56 p.m. Calls were sent to the navy and the coast guard to send firefighting tugs to assist the two Boston fireboats.

The fire had gained tremendous headway, and Boston firefighters were unable to enter the structure to fight the blaze because of the lightning speed with which it spread. Firefighters attacked the blaze with deck guns, hand lines and a water tower on Commercial Street, from the cobblestone yard within the wharf, from flanking wharves and from a flotilla of fireboats in the harbor. Heavy-stream appliances were directed to protect a five-story warehouse to the north on the same wharf. The raging inferno of rolling flames and heavy, black smoke threatened the entire waterfront.

Twenty-nine engine companies, five ladder companies, three rescue companies, two water towers and two fireboats fought the fire. Units from Brookline, Cambridge, Somerville, Chelsea and Everett, as well as several firefighting tugs, responded directly to the scene to assist Boston.

A public service announcement on Boston radio stations urged citizens to be particularly careful with fire and to assist by fighting grass and rubbish fires themselves, since the resources of the Boston Fire Department were stretched thin.

Veteran firefighters reported at the time that they could see no farther than three feet in front of them due to the extremely heavy smoke caused by the burning rubber, leather and creosote. Firefighters held their positions despite the punishing conditions. Over one hundred firefighters and police officers were treated for smoke inhalation, and fifteen firefighters were hospitalized. Chief of Department John V. Stapleton reported that the department record for amount of hose used at a single fire was broken that day.

Throughout the afternoon and into the evening, the firefighters kept up their epic battle before a crowd of 25,000 spectators. At 5:00 p.m., Stapleton ordered all members to remain on duty beyond their 6:00 p.m. relief time, effectively doubling the force during the emergency.

By nightfall, the shed and wharf were near collapse. By 9:30 p.m., the fire was contained and companies began returning to quarters, but a major complement of Boston apparatus would be kept on the scene for several days until the extinguishment was complete.

Only the highest degree of efficiency on the part of all services, wise leadership by the chief officers and the devotion to duty by the men kept the fire from spreading far beyond its point of origin.

Today, Union Wharf is occupied by waterfront condominiums. There is little sign of the sprawling warehouse or the heroic struggle made by firefighters that Sunday afternoon to save the Boston waterfront so many years ago.
[P.C.]

7. North End Firehouses

Beginning with Engine 8 site, 392 Hanover Street, North End
While many of Boston's firehouses have disappeared over time as companies disbanded or consolidated, the city has a good number of historic firehouses remaining. The Engine Company 8 and Ladder Company 1 firehouse in the North End was built in 1948 on Hanover Street at the corner of Charter Street. This was the third firehouse occupied by Engine 8. It has operated as a regular company since it was organized on North Bennett Street with a horse-drawn steam fire engine on November 1, 1859. The Hanover Street firehouse opened September 15, 1948, when Engine 8 moved here from older quarters on nearby Salem Street. Ladder 1 moved September 22 from its temporary quarters at the

Standing in front of Engine 8's ice-encrusted hose wagon, Chief of Department Peter E. Walsh monitors the progress of a three-alarm fire near North Station on January 18, 1920. Walsh served as chief from July 31, 1919, until his retirement on March 3, 1922.

Members of Engine Company 8 stand in front of their quarters at 133 Salem Street with their horse-drawn apparatus, circa 1915. This station served as Engine 8's second firehouse from 1868 to 1948, when a new station (still in use) was built at 392 Hanover Street.

Bowdoin Square firehouse. Ladder 1's quarters were originally located on Friend Street near Haymarket Square for many years. The company relocated to Bowdoin Square in 1933 when the firehouse was demolished for the widening of Merrimac Street.

Walk up Hanover Street two blocks and turn right into North Bennett Street. On the left at the corner of Bennett Place at number 12 is a four-story apartment building; in 1859, this was the original firehouse for Engine 8. This structure was built as a two-story brick dwelling house about 1800. In the mid-nineteenth century, the neighborhood declined and most of the houses were converted to rooming houses. Number 12 was purchased by the city in 1859 and rebuilt for use as a firehouse. The building was extensively remodeled, and a stable for two horses was added in the rear, as well as sleeping quarters on the second floor for the driver. When the city vacated the building in 1868, the structure was sold to a private owner who operated a livery stable in the building. In 1882, the building was sold to a North End businessman who remodeled the building again, this time adding two floors and converting it to the apartment building that you see today.

Continue up North Bennett Street and turn left on Salem Street. Just past Prince Street on the left is Sheldon's Market at 133 Salem Street. In 1868, the city built a new

firehouse at this site for Hose Company 1, which had previously occupied an old two-story brick building that is still standing at the corner of Carroll Place, later changed to Jerusalem Place. The new building was modern in every way for its time, and when the new house was finished, the Boston Fire Department decided to move Engine 8 in with Hose 1 due to the difficulty of turning and backing an engine in narrow North Bennett Street. The firehouse at 133 Salem Street opened in January 1869 and was in continuous use almost eighty years. Hose Company 1 was disbanded in 1873. In 1917, when Engine 8 was converted from horse-drawn to motor apparatus, the firehouse went through a thorough renovation. The stables were removed, the main floor was concreted and an entirely new front was added, dramatically changing the appearance of the building. By the 1940s, with the difficultly of navigating traffic on narrow Salem Street, the city built the current firehouse at Hanover and Charter Streets. The building at 133 Salem Street now is a retail store with apartments overhead.
[J.T.]

8. The Paul Revere House

9 North Square, North End

Most Americans know Paul Revere as the man on horseback who rode from Boston toward Lexington to warn Revolutionaries that the British were coming. Others may know he was a fine and exacting copper- and silversmith. But Revere (1734–1818) holds a special place in Boston as a volunteer firefighter and fire warden, a group that also includes John Hancock and Sam Adams. Indeed, to be a volunteer firefighter in the colonial period was deemed a patriotic duty. Boston had the nation's first paid fire department, dating back to 1678, when Thomas Atkins was designated foreman of a company that would handle a newfangled fire "masheen" brought from England. In 1711, after another devastating fire, ten fire wards were chosen; they were given the authority to organize firefighting efforts and arrest looters. In 1718, citizens organized the Boston Fire Society, a mutual aid organization. By 1760, the city had nine fire companies and the city was divided into fire districts. For many prominent Bostonians, it was a badge of honor to be a volunteer member of a particular fire company. John Hancock, who was a fire warden along with Samuel Adams, gave a hand-pumped fire engine to Boston; it was named Hancock Number 10 in his honor, according to fire historian Paul Ditzel. Revere was appointed a fire warden in 1775, after his famous ride, according to Ditzel.

A final note: one of Revere's apprentices was coppersmith William Cooper Hunneman. The company he founded went on to build some of the region's finest hand-pumped fire engines. The Hunneman Company, which had a manufacturing plant on Hunneman Street in Roxbury between Washington Street and Harrison Avenue, also made hose reels and ladder trucks; it was in business from about 1802 until 1883, when the newer steam-powered engines made hand-powered engines obsolete. Today, Hunneman hand-pumped engines are prized collectors' items among those interested in antique fire machinery. Revere's apprentices also included Ephraim Thayer, another fire engine builder. Thayer's first fire engine is on display at the Boston Fire Museum. (See entry on 344 Congress Street fire station.)

[S.S.]

9. Merrimac Street Fire

116–126 Merrimac Street, North End

In the early morning hours of February 5, 1898, a Boston Police patrolman saw smoke coming from the upper floors of the building occupied by the George Bent Bedding Company at 116–126 Merrimac Street. At about 3:55 a.m., he pulled the alarm at Box 412, Causeway and Lowell Streets. By the time the first fire company arrived, smoke and fire were pouring from the upper levels. Assistant Chief William T. Cheswell ordered a second alarm transmitted at 4:13 a.m. when he arrived. "Run hose lines into the building by the stairs. Get lines over the ground ladders," he yelled to the men. A very heavy snowstorm a few days before had left hydrants covered with ice and snow; some hydrants were even frozen. Firefighters frantically tried to get water on the fire. Within three minutes, a third alarm was ordered.

The Merrimac Street building, built in 1836 for the Union Baptist organization, was five stories high and about 125 feet wide and 75 feet deep. It was loaded with large amounts of feathers, excelsior and other materials used in the manufacturing of furniture and mattresses.

Among those responding on the second alarm were District Chief John F. Egan from Fort Hill Square, Captain Joseph M. Garrity and Lieutenant John J. O'Connor of Engine Company 7 from East Street. "Egan! Go to the fourth floor with Engine 7," Cheswell ordered. Engine Company 39 had run lines to the fourth floor and Egan, a dashing man known for his bravery, took command of this area. His last order was yelled from a fourth-floor window. "Bring me up a lantern," he cried before his white helmet faded back into the room out of sight. His driver was crossing the street with a lantern when tragedy struck.

At about 5:15 a.m., when the fire was just about under control, the roof collapsed into the fifth floor and all floors came tumbling down, trapping firefighters beneath tons of debris. Frantically, firefighters began to dig in the debris to rescue their comrades; it was tough work because of the heavy fire load and the snow. Lieutenant O'Connor had just spoken with District Chief Egan and knew about where he should have been. District Chief Peter McDonough and District Chief John Ryan were placed in charge of rescue work. So many firefighters volunteered help that they had to wait outside the area to give others room to work.

Fire Commissioner Henry S. Russell sent for heavy equipment and stevedores to remove the heavy timbers. From behind the pile, Captain Garrity called, "We're way down here, boys. For God's sake be as quick as possible."

As the men dug, Garrity and Hosemen Thomas Conroy, Edward Shea and Phillip Doherty called out to guide rescuers to where they were trapped. Hoseman Shea of

The victims of the Merrimack Street Fire of February 5, 1898. *Clockwise from upper left*: Captain James H. Victory, E 38–39; District Chief John F. Egan, District 3; Lieutenant George J. Gottwald, E38–39; Hoseman Patrick H. Disken, E38–39; Hoseman John H Mulhern, E38–39; and Hoseman William Welch, E38–39.

Engine Company 7 was the first firefighter found; he was discovered about forty feet from the front of the building, buried under a large pile of bedsteads. Next was Captain Garrity, who was severely injured. Just after 7:00 a.m., Hosemen Conroy and Doherty, both from Engine Company 7, were located; it took another hour to free them all and send them to Massachusetts General Hospital for treatment. When Conroy was being carried from the building, he spotted Fire Commissioner Russell and saluted as he was carried by. "Do you hear any more voices?" District Chief Ryan asked the rescuers. None was heard, yet firefighters continued to dig. District Chief Egan was found about 10:00 a.m., and right behind him was Captain James Victory of Engine Company 38–39. Egan was found in a sitting position with his arms in front of him, like he was trying to fight his way through the huge pile of wet feathers. Both men were alive and were given their last rites by Father Finnegan of St. Mary's Church in the North End. They were rushed to Massachusetts General Hospital in an ambulance, but they both died before they arrived. When Egan was carried out of the building, many recognized him and wept upon seeing his condition.

Colonel Henry S. Russell was appointed commissioner of the Boston Fire Department on July 1, 1895. His appointment as a single commissioner replaced the multimember Board of Fire Commissioners, which had been in place since November 20, 1873.

Hosemen Disken and John H. Mulhern of Engine Company 38–39, both dead, were found about 10:10 a.m. The helmet of Lieutenant George J. Gottwald was found and then Hoseman William Welch and Gottwald, the last men, were discovered. Gottwald had not been seriously injured, but had suffocated by being buried with wet feathers.

Arguments raged over the cause of the fire and why the building had collapsed. There had been other fires in this building a few years previous and Building Commissioner John S. Damrell had stated that repairs had been made. Others were not so sure. Superintendent Samuel Abbott of the Protective Department said he would not have sent his men inside, as it was a "dangerous building." Chief of Department Lewis P. Webber insisted that "at the time of the falling of the floors there had at no time been enough fire on either floor to weaken the supports to such an extent as to make the building dangerous or the crash came without warning."

Whatever the cause, six firefighters were dead and a number of others were injured. Funds were started for the members' families by two Boston newspapers; more than $50,000 was raised, which was a lot of money in 1898.
[B.N.]

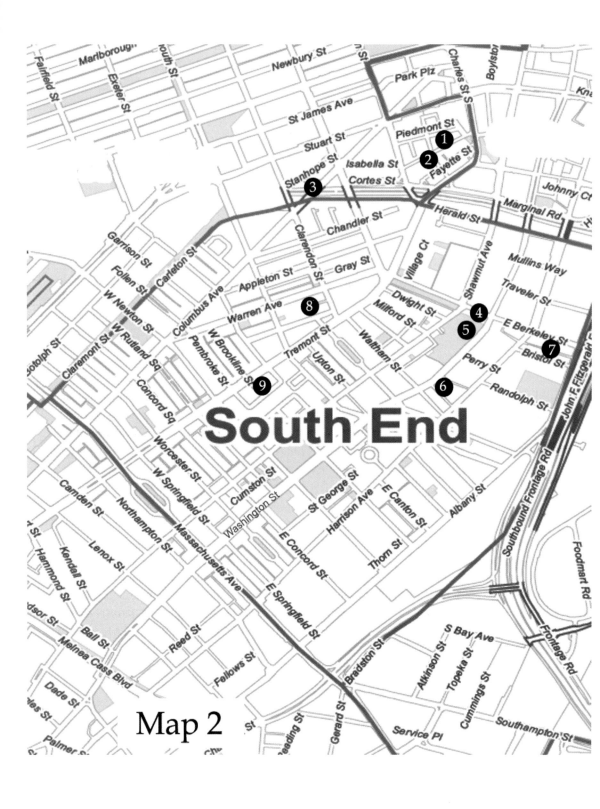

Map 2

Map 2.

South End

1. The Cocoanut Grove Nightclub Fire

15–17 Piedmont Street, South End

One of the worst fires in American history occurred on a chilly night on November 28, 1942, in Boston's popular Cocoanut Grove nightclub. The club, licensed for five hundred occupants, was packed that night with more than one thousand people, including sailors, soldiers, college football fans and people celebrating birthdays, anniversaries and weddings. Decorated with a South Seas theme, the club was one of the most popular hot spots in Boston, featuring dining, dancing, singers and a band. It had a main dining area, a downstairs bar called the Melody Lounge and a newly built lounge that opened onto Broadway. Earlier that day, the heavily favored Boston College football team had gone down to defeat to Holy Cross 55–12. BC officials sadly canceled plans for a victory party that had been previously scheduled for that night at the Grove, but football fans flocked to the nightspot, some to celebrate, others to drown their sorrows.

About 10:15 p.m. in the downstairs Melody Lounge, a busboy lit a match to see where to put in a light bulb. Moments later, the cloth ceiling of the bar caught fire and within seconds a huge fireball roared through the entire club.

Many people ran to get out through the main entrance of the club, a revolving door. It jammed from the pressure of the crowd, trapping people inside. Other patrons raced for other exits, but many had been locked from the inside in an effort to keep people from running out on their bills. The fire, moving with ferocious speed, trapped hundreds of people in the inferno.

By coincidence, firefighters happened to be a block from the club when the fire broke out. A couple had spotted a smoldering fire in the seat of an automobile and one of them pulled the alarm at Box 1514, Stuart and Carver Streets. Because the box was in a high-density area, eight companies responded: Engine Companies 7, 10, 22 and 35, Ladder Companies 13 and 17, Rescue Company 1 and Water Tower 2. Division 1 Deputy Chief Louis C.I. Stickel and District 5 Chief Daniel Crowley also arrived. As the companies were being dismissed after putting out the automobile fire, firefighters

During the April 17, 1928 dedication of the new fire station at 194 Broadway in the South End, apparatus was impressively displayed on the apron of the station. This station closed on May 10, 1971.

heard the yell, "There's a fire at the Cocoanut Grove!" Some firefighters ran down the street; others jumped on the engines and drove in the direction of the club.

For many club-goers, it was too late. People were pouring out of the club, some collapsing on the sidewalk. Others were trying to get out from the revolving doors on Piedmont Street—that area was so hot that firefighters could not get close enough for rescues, and were forced to watch people burn to death even as water was being poured on the doors. However, firefighters were able to make rescues immediately via the exits on the Shawmut Street side of the main dining room. Crowley ordered a first alarm, which was sounded from Box 1521 at 10:21 p.m. Skipping the second alarm, Deputy Chief John "Black Jack" McDonough ordered a third alarm, just as Stickel ordered a second and third alarm.

Firefighters worked on three sides of the club. Engine 7 got a line on the front entrance at Piedmont Street, while other companies worked on the other side of the club on Shawmut Street. Others broke through block-glass windows on Broadway to reach the new lounge. Some firefighters sustained life-altering injuries trying to get people out of the raging fire. Some used axes to try to break down an emergency door that had been bolted from the inside. A pile of bodies was found inside behind this door.

Among those carried out was Buck Jones, a famous cowboy star who was in Boston on a promotional tour. Suffering from a cold, he had not even wanted to go to the club that

Engine 35's self-propelled 1898 Amoskeag steamer responded to a call from its quarters on Mason Street, downtown. A rough ride was ensured, due to the hard rubber tires, cobblestone streets and streetcar tracks. This steamer served at Engine 35 from January 1898 to January 1921.

The Cocoanut Grove nightclub was the premier entertainment venue in the city in the late 1930s and early 1940s. After the fire had been extinguished, large crowds gathered to inspect the ruins, although visible evidence does not indicate the horror that occurred inside.

night. He died of his injuries. The dead included an entire wedding party, four brothers from the same family and members of the Cocoanut Grove staff who heroically stayed behind to try to guide people to safety. Victims were rushed to local hospitals, including Massachusetts General Hospital and Boston City Hospital. Boston City Hospital received most of the victims; at one point a patient was arriving every eleven seconds.

Many victims suffered terrible burns, while others were felled by toxic smoke from the burning décor. Some, with seemingly healthy cherry red complexions, suffered from carbon monoxide poisoning, while others suffocated as the fire consumed all the oxygen in the club. Victims rushed to Boston hospitals were treated with penicillin—the first time it was used on a civilian population—and new techniques for treating burns and lung injuries were tried out.

By 10:45 p.m., the main body of the fire was out and firefighters were tasked with the grim chore of removing bodies. Some victims were found sitting at tables, felled where they sat.

The official death toll was put at 492—including one man who had escaped from the fire, but felt so guilty that his wife had died in the inferno that he jumped out a hospital window and committed suicide. More than 160 people were injured. No firefighters

The removal of victims from the Cocoanut Grove was a strenuous task. Many uninjured patrons and bystanders, including navy and coast guard personnel, aided in this effort. In this view, firefighters pass an injured party out to waiting stretcher-bearers.

were killed, but many were haunted for the rest of their lives by the events of the night of November 28.

In the days after the fire, a huge investigation was launched as to its cause. To this day, many wonder why the fire moved so quickly through the club. The busboy who had lit the match was exonerated. Officially, the cause of the Cocoanut Grove fire is listed as being "of unknown origin."

More evident was the club's poor safety record. The club had been inspected just ten days before the fire. However, a hearing and a trial found that the owners had used unlicensed electrical contractors and that many of the furnishings were not treated with flame retardants. Only Barnett Welansky, the owner of the club, was convicted of manslaughter in connection with the fire and sent to prison.

The impact of the Cocoanut Grove fire reached around the nation as communities passed (or began to enforce) safety codes with regard to emergency exits. Today, all revolving doors in the state must have regular outward-swinging doors nearby. Additionally, the fire led to innovations in burn treatment that are now common today. Legal history was made in the prosecution of the fire, and Cocoanut Grove precedents have been used in other cases in which building owners have not maintained safe conditions.

Today, the streets in this section of Boston have been altered. The block where the "new" Broadway lounge was located is now a parking lot and garage. The area of the main dining room is now bisected by a new street. There is a sign about the fire on the side of the garage; in the sidewalk on Piedmont Street, near where the revolving door was located, is a bronze plaque marking the tragedy. It was installed in 1992 to commemorate the fifty-year anniversary of the fire, and made by a club employee who had survived the fire. Flowers are still left at the site by those who never want the lessons of the Cocoanut Grove to be forgotten.

[S.S.]

2. Chemical 2 Quarters

25 Church Street, Bay Village, South End
This small brick firehouse was built in 1869 for the horse-drawn Hose Company 8, which moved to this area of Boston—then called the Church Street District—in April of that year. The area, now called Bay Village, looks now very much as it did back then: a neighborhood of small brick homes. On May 7, 1874, Hose Company 8 was disbanded and replaced by Chemical Company 2; the company's designation is still engraved in the granite lintel of the building above what was once the apparatus bay.

Chemical wagons were innovative additions to the fire service in the 1870s and served many purposes. Boston pioneered the use of chemical companies, which were first called extinguisher wagons. These were open wagons carrying about a dozen soda and acid fire extinguishers that were used to control fire until larger hose lines could be connected to hydrants and steam fire engines. The "extinguishers" did not put chemicals on fires; rather, the action of the soda and acid created pressure to allow streams of water to be played on blazes. Boston had three extinguisher wagons, two of which evolved into much more advanced chemical engine companies equipped with larger soda and acid–activated water tanks, which could play steadier and larger volumes of water on fire. Over time, fourteen such chemical engine companies served the fire department. Chemical 2 remained in service until 1920, when it was disbanded as chemical tanks became standard equipment on motor hose wagons, thus making chemical companies redundant. Several companies, including Engine Company 35 and Rescue Company 1, were temporarily housed here until the firehouse was permanently closed in 1928. The building was sold and converted to residential use.

After decades of mixed use, which included nightclubs, bars and motion picture warehouses, Bay Village experienced a housing revival as new homeowners began to renovate their nineteenth-century properties. The downtown location attracted young professionals who infused the old district with new life. Today, the area is thirty years into its rejuvenation and remains a highly desirable location. Despite all the changes in the neighborhood, the little chemical house still stands at 25 Church Street.

[T.G.]

This station was built in 1869 to house Hose Company 8. Chemical Company 2 was organized here on May 7, 1874, and remained in service until July 20, 1920. The building was sold in 1928 for residential use and is still extant.

3. Pope Bicycle Building Fire

221 Columbus Avenue, South End

In the 1890s, bicycles were the rage all across America. All classes of people, including the wealthy, owned "wheels." On spring and summer weekends, young men and women rode their wheels into the country to visit familiar spots such as Chestnut Hill Reservoir and Franklin Park. The bicycle trade was prosperous, with as many as fifty thousand units being sold per year in the Boston area. The heart of this industry was the so-called Bicycle Row on Columbus Avenue. Dozens of firms had showrooms along this street, the most famous being the Pope Bicycle Company, located in a handsome five-story building at 221 Columbus Avenue near Berkeley Street. This white brick building with terra cotta trim was built in 1891 and contained showrooms with hundreds of bikes, and on the top floor there was a riding school for wealthy Back Bay clientele, with ten instructors on duty to train novices in the art of bicycle riding.

This stately building of white brick and terra cotta trim was gutted in a fire on March 12, 1896. The Pope Bicycle Factory contained bicycle showrooms on the lower floors and a riding academy on the top floor to accommodate the bicycle rage of the Gay Nineties.

On Thursday, March 12, 1896, fire started in the basement of the Pope building at about 3:30 p.m. and spread rapidly through the structure. The basement was full of spare bicycle parts and packing crates and it is thought that these were the origin of the fire. Salesmen on the first floor detected an odor of smoke and then saw smoke coming through cracks in the floor, which grew steadily worse. Maintenance men at work in the building picked up fire extinguishers and started for the basement, but were driven back when they reached the cellar stairs. A building fire alarm was pulled and the bells notified occupants in the upper floors to escape. Office workers and customers at the riding school on the top floor were brought to the street by elevator, and most of the occupants escaped quickly.

Box 96 at Columbus Avenue and Berkeley Street was sounded at 3:36 p.m. Firemen arriving on the first alarm ran their lines into the first floor, only to be driven back by heavy fire. Some employees had to be taken out by ladder from the second floor, as their escape by the stairway was now impossible. The floors in the building were finished in pine and there were no partitions on the lower floors to retard the fire as it spread upwardly with remarkable speed. The second alarm was skipped and the third alarm was transmitted at 3:45 p.m. The fourth alarm was ordered at 3:53 p.m. and the fifth alarm at 4:08 p.m. Arriving firefighters went into an offensive attack, running lines into the adjoining Youths Companion building, and played water out the windows from every floor into the fire building. Deck guns were set up in the front, and the water tower that was called to the fire operated from Columbus Avenue on the fire, which by then was a raging furnace. Linemen climbed poles in front of the building to cut the electric wires and two of them received shocks. These were the only serious injuries at the fire. The fire drew a large crowd of weekday sightseers, and service on the electric cars had to be stopped. Albert Pope and his staff were seen during the fire in an adjoining store as Pope negotiated to set up a temporary location to allow the company to remain in business. At 5:00 p.m., Pope announced that he had secured the rent of a store at 200 Boylston Street and would be open for business at that location the following morning. Pope had wired to the firm's factory in Hartford, instructing them to ship three hundred bicycles to Boston for business the next day.

The main fire was under control by about 6:30 p.m., and the building was a mass of ruins with only the four walls still standing. Hundreds of bicycles were destroyed, and the total loss amounted to $275,000. The all out on Box 96 was sent at 11:30 p.m.

The following year, a new seven-story building was built by the Pope Company at the same location, and that building is still standing today. The building later became home of the trade magazine publishing firm Cahner's Publishing Company. It is now the home of the popular Mistral Restaurant on the first floor, with commercial office space on the upper floors. If you look closely, you can see the letters P-O-P-E engraved across the top of the building and the year 1897 engraved on the front. The neighboring Youths Companion building also still stands.

[J.T.]

The Roosevelt Hotel fire occurred at 3:45 a.m. on February 4, 1968, in a transient hotel on Washington Street. Occupants were rescued in freezing temperatures over fire department ladders.

4. Roosevelt Hotel Fire

1147 Washington Street, near East Berkeley Street, South End

On February 4, 1968, the Fire Alarm Office received a hasty and incomplete phone call from the Hotel Roosevelt. Fire Alarm operators dispatched Engine Company 56 (covering Engine Company 3) and Ladder Company 17 to this South End hotel. This was followed by the striking of Box 1632 at 3:49 a.m. Upon arrival, Lieutenant Donald Toomey of Ladder 17 discovered a fast-moving fire roaring up the straight open stairwell of the old building. Firefighters could also see people trapped on the upper floors. Toomey ordered a second alarm and put his men to work with extension ladders from the street and from the elevated railway platform of the Dover Street Station.

Also responding was District Four Fire Chief Leo D. Stapleton, who ordered a third alarm at 3:54. a.m. (Stapleton would later become fire commissioner.) Eventually, five alarms were struck. The elevated railway trestle nearly filled this narrow portion of Washington Street, reaching from sidewalk to sidewalk, and firefighters had great difficulty in working around this obstacle. Due to this obstruction, only one aerial ladder could be deployed and then only at one corner of the building. Tragically, nine people were killed, but due to the skill of Boston firefighters more than one hundred hotel residents were rescued over ladders that night. Accordingly, the fire commissioner commended all personnel who fought this fire. Soon afterward the hotel, which had seen many fires over the years, was torn down. Today, the site is partially taken up by Peters Park in a neighborhood undergoing upscale transition.
[B.M., T.G.]

5. Briggs Place Fire

7–11 Medford Court and 1217 Washington Street between Washington Street and Shawmut Avenue, South End

This December 9, 1965 fire, which swept through the block-long, four-story warehouse of the Boston Leather and Rubber Company with lightning speed, is noteworthy because of a wall collapse on the high-pressure hose wagon of Engine Company 26 at the height of the blaze. The engine was severely damaged, but it continued to operate. A dramatic photo was snapped of the apparatus covered with bricks as the deck guns continued to discharge water. The photo was highly publicized in advertisements as portraying the indestructibility of Mack fire apparatus. When the bricks were cleared, the wagon was driven away under its own power. The body was rebuilt at the Fire Department Shops and returned to service. It continued in service with Engine 26 and then at Engine 7 until 1975. Twenty-one engine companies and five ladder companies responded to this five-alarm fire at Box 1614; there were no serious injuries and no fatalities. The area is now occupied by Peters Park.
[B.M., T.G.]

A fire in this five-hundred-foot-long, five-story building on December 9, 1965, on Briggs Place in the South End caused a collapse of a brick wall onto the Mack hose wagon of Engine 26. Despite the damage, 26's deck gun continued to operate.

6. Arcadia Lodging House

1202 Washington Street, South End

At the turn of the century, the South End was a busy, congested area with numerous tenements and lodging houses. Laconia Street was located only a couple of blocks from the busy Dover Street area, with its station for the old Boston Elevated Railway, better known as "the EL." Among the so-called "flophouses" was the Arcadia Lodging House on Washington Street at the corner of Laconia Street. The five-story brick building had a mansard roof, a spiral staircase in the front and a balcony fire escape in the rear and could accommodate 243 men. On December 2, 1913, 159 men had registered, paying fifteen, twenty or twenty-five cents for a bed. About 2:00 a.m. on December 3, a fire started in a first-floor closet. The blaze may have been an accident, or it may have been set intentionally by a man who was refused admittance earlier in the evening. Eighteen-year-old Warren Crowell, who was sitting in the reading room as he did not have money for a room, discovered the fire. He alerted the night clerk, James Welsh, and the night watchman, Arthur McGlynn, who started to wake residents. McGlynn ran down the stairs, through the flames and sounded the alarm at 2:04 a.m. at Box 771 at Washington

Street and Cottage Place. But McGlynn, in his hurry, failed to ring the house alarm and shut the door behind him—actions he would come to regret deeply.

By the time Engine Company 3 and Ladder Company 3 arrived, the building was fully involved, with residents crying for help at windows. At least two had already jumped to their deaths. The only escape was by firefighter ladders, as there were no outside fire escapes. Several residents went to the roof and managed to cross over to the building next door. The final death toll was twenty-eight, of which eighteen were burned beyond recognition and could not be identified. The local Elks Lodge Number 10 offered to pay for the burial of the unknown victims. On February 8, 1914, the unidentified victims of the Arcadia Lodging House were buried at Mount Hope Cemetery in Roslindale. A properly engraved granite stone was placed at the site.

[B.N.]

7. Historic Boston Fire Headquarters, 1895–1951

60 Bristol Street, South End (Now Paul Sullivan Way)
Bristol Street has been associated with the fire department for many years because the headquarters complex, a familiar landmark in the South End, was located here from July 1, 1895, to August 6, 1951. The distinctive six-story brick building and the adjoining eight-story tower were modeled after the Palazzo Vecchio in Florence, Italy. Before the adoption of radio communications, a flashing signal light atop the distinctive tower was used to signal fireboats whether to continue responding up the Fort Point Channel to alarms in the Albany Street/South Bay area—which required the opening of many drawbridges—or to return to quarters if not needed. A familiar and breathtaking sight in the South End was that of the new firefighter recruits scaling the eight stories using single-beam pompier ladders as part of their training. The adjacent fire department repair shop building at the corner of Albany and Bristol Streets had been located there since 1885.

The Boston Fire Department first established a facility in the area in 1875, when it built a firehouse at 440 Harrison Avenue, on the corner of Bristol Street, for Engine Company 3 and Ladder Company 3. The firehouse remained at this location until 1937, when structural problems forced the fire department to seek a new location. A new firehouse was completed in 1941 at Harrison Avenue and Wareham Street.

In the 1880s, the fire department proposed to build a new repair shop. The old shop at Harrison Avenue and Wareham Street was cramped and not properly equipped for the maintenance of a large fleet of fire apparatus. A vacant lot of land was purchased by the city at Albany and Bristol Streets and a modern three-story brick shop building was erected. It contained spacious rooms and tools and machinery needed for the building and repair of fire apparatus. The new shop opened July 11, 1885.

As the years went by, discussions had been held concerning the need of a fire headquarters building to consolidate the many functions of the fire department. This chance came after the disastrous Thanksgiving fire of November 28, 1889, when the

This building was the headquarters of the Boston Fire Department from 1895 to 1951. The drill tower was modeled after the Palazzo Vecchio in Florence, Italy. The building is still extant and is now the Pine Street Inn homeless shelter.

The Fire Department Repair Shop in ruins after a fire on August 9, 1910, which also destroyed the Blacker and Shepard lumberyard. The drill tower at Fire Department Headquarters can be seen looming over the ruins.

city council appropriated $510,000 for the purchase of new fire apparatus and stations. Included in this total was $165,000 that was intended for building a fire department headquarters building. Due to the pressure of real estate interests in the 1890s with rapid development and high prices, the city was unable to find an adequate site at a reasonable cost to meet this need, and the project languished for several years. After much negotiation, the Boston Fire Department selected a city-owned lot at 60 Bristol Street, adjacent to the recently erected shop building. The city architect prepared plans for the elaborate new edifice, which was completed in December of 1893.

The first unit to occupy the new building was Water Tower Company 2. This apparatus was purchased in 1890 and held in reserve for three years, as there was no room to place it in service in a firehouse. The new building on Bristol Street had space on the first floor for apparatus, and Water Tower Company 2 entered service on December 28, 1893. Other fire department functions began moving into the new headquarters building in February of 1894. It was not until July of 1895 that the new headquarters was officially opened. Also moving to Bristol Street was the Fire Alarm division. The Fire Alarm Office moved its office from its old location at city hall on School Street to the top floor at Bristol Street on May 20, 1895.

This is an interior view of the Boston Fire Department Shop located at 363 Albany Street, which was partially destroyed as a result of the Blacker and Shepard lumberyard fire on August 9, 1910.

Although the new fire headquarters was the pride of the department, there was some criticism of the location, due to the hazardous nature of the neighborhood with its congestion and fire hazard potential. The critics were proved right on August 10, 1910, when a major fire started in the lumberyards of the Blacker and Shepard Lumber Company on the east side of Albany Street, just south of Dover Street and next to Fort Point Channel. The site consisted of piers, sheds and two- and three-story wooden buildings used for lumber storage, all of which quickly became involved. The fire spread out of control and destroyed many businesses on both sides of Albany Street. Firefighters were driven back as the intense heat forced them away from their positions, and the radiated heat ignited the Fire Department Repair Shop on the west side of Albany Street. Fire spread rapidly through the shop building, and many fire engines were quickly moved to safety to the yard outside the building. Several important engines were lost, however, including Engine 9, Chemical 5, Ladder 23 and the aerial ladder of Ladder 14. The building was a total loss. Fire also threatened the headquarters building next door, but it was saved because the outside sprinkler system was activated. The Fire Alarm Office was ready to evacuate at a minute's notice, but this was not necessary.

The fire renewed criticism about the location of Fire Alarm Headquarters so close to a hazardous industrial area. It would be another fifteen years before a new location for the Fire Alarm Office was secured at 59 The Fenway. Following the fire in 1910, a new repair shop was built on the same site, and the new building was completed in July of 1911. A completely new fire headquarters and repair shop complex was built at 115 Southampton Street in Roxbury and opened for business on August 6, 1951.

The headquarters building, drill tower and repair shop still stand on Bristol Street. The property has been the site of the Pine Street Inn homeless shelter since 1978. [J.T.]

8. Engine 22 Historic Firehouse

70 Warren Avenue, South End

The attractive, three-story, two-bay former firehouse on Warren Avenue near Clarendon Street dates from 1901 and was quarters for Engine Company 22 and Ladder Company 13 until it was closed in 1960. Engine Company 22 was organized on October 14, 1873, at a temporary location on Parker Street (now Hemenway Street) near the railroad crossing at Boylston Street. The company then moved to a new firehouse that was located at the railroad crossing on Dartmouth Street in the Back Bay on July 3, 1875. When the New Haven Railroad decided to build its new Back Bay passenger station, it was necessary to move Engine 22 and demolish the firehouse property for the railroad development. The old firehouse closed on May 9, 1899, and Engine 22 temporarily relocated to fire headquarters on Bristol Street. After prolonged negotiations, the city purchased a vacant lot of land on Warren Avenue in the South End to build a new firehouse.

The fire department had also been looking for some time to build new quarters for Ladder Company 13, located at 1157 Washington near Dover Street. This ancient and very small building was built in 1837 for a hand-drawn volunteer engine and was entirely inadequate for a horse-drawn aerial ladder truck.

The new building on Warren Avenue was finished, and Engine 22 and Ladder 13 moved to the new firehouse on August 1, 1901. This firehouse was also the headquarters for Fire District 7 until 1954, when through reorganization it became District 4. It was also the quarters of the deputy chief of Divisions 1, 2 and the citywide deputy chief at various times.

For many years, Engine 22 was one of the city's busiest fire companies, as it was located in a neighborhood filled with rooming houses and other buildings prone to fire. The Warren Avenue firehouse is also significant because for a time it housed the first automobile fire apparatus in Boston: a 1906 American LaFrance chemical car, mounted on a modified Packard touring car chassis, which was loaned to the fire department. It was designated Auto Chemical 13 and was under the charge of the captain of Ladder Company 13.

In 1959, work began on a new firehouse at 700 Tremont Street at the corner of Concord Street. Engine 22 and Ladder 13 were moved to that location on June 11, 1960. The firehouse on Warren Avenue remained vacant for several years; then, exactly two years after it was vacated, the building suffered major damage to the top floor in a four-alarm fire on the rain-soaked night of June 11, 1962. Engine 22 remains in service

Built in 1901, this firehouse served as quarters for two busy companies, Engine Company 22 and Ladder Company 13, from August 1901 until June 1960.

on Tremont Street, but Ladder 13 was deactivated on October 20, 1981. The building sat vacant for several more years, but was renovated in the 1970s and used by the Boston Redevelopment Authority as offices. It was later sold and converted to its current use as condominiums.

[J.T.]

9. The Trumbull Street Fire

West Brookline Street and Shawmut Avenue, South End
Trumbull Street was a very small street located in the South End off Newland Street near East Brookline Street and Shawmut Avenue; it was only wide enough to allow one automobile to pass. On the night of October 1, 1964, at about 12:30 a.m., a passerby spotted a fire on Trumbull Street and pulled Box 1671, located at Shawmut Avenue and West Brookline Street. Engine Company 3, with Lieutenant Christopher "Steve" Fraser in command, and Ladder Company 3, with Lieutenant John Campbell in command, responded. Filling out the assignment were Engines 22 and 7 and Ladder 13. As the fire engines neared the box, the men could see a glow in the sky; the passerby told them the location of the fire and the companies began a circuitous route through the narrow streets to the building. Only the hose wagon of Engine 3 was able to maneuver into Trumbull Street. Ladder Company 3 could only reach the corner of Trumbull Street.

Heavy fire was observed in the upper floors of a vacant four-story factory at 34 Trumbull. Campbell reported "fire showing" and then ordered a "working fire" at 12:36 a.m. Engine Company 3 ran a big line into the building, and firefighters attempted to make their way to the second floor, but were forced back to the landing by the heavy fire. Ladder Company 3 raised ground ladders and the one-hundred-foot aerial was raised to the roof of an adjoining building. All the ladders of Ladder 3 were utilized.

Deputy Chief Frederick P. Clauss did not like the look of the conditions in this one-hundred-year-old building, and he ordered everyone out. He repeated this order to District Fire Chief John McCarthy and company officers, and then meticulously verified that the order was being carried out.

Engine Company 24, under the command of Lieutenant John J. McCorkle, took a big line over one of the two thirty-five-foot ladders placed in the front of the building while Firefighter James B. Sheedy of Ladder Company 4 was preparing to "dog" the other ladder. (Dogging is a procedure for securing portable extension ladders at windows or parapets; a "dog" is a length of chain with a spike at one end and a loop at the other that is used to keep the ladder in place.) Lieutenant John J. Geswell of Ladder Company 26 (detailed to Ladder Company 4) said he would dog the second ladder in preparation for another line of hose. As he was passing Firefighter Frank Enrici of Engine Company 24, he cautioned him that the ladder had not been yet dogged. Those were his last words.

Without warning, part of the building collapsed. Men were knocked from ladders and the balcony fire escape, while other men on the ground were buried. The firefighters from Ladder 3, working in the rear of the building, heard the loud crash and ran to help. All

This photo shows companies attacking a five-alarm fire at 36 Trumbull Street on October 1, 1964. This photo was taken prior to the building's collapse, which claimed the lives of five Boston firefighters

they could see was a large cloud of dust. As they moved in to try to rescue the wounded, the rest of the building collapsed, burying all the men in piles of masonry and dust. Clauss was knocked to the ground and, though injured, he managed to call to Captain Leo Wisentaner of Ladder Company 15: "Leo, strike a fifth alarm and get ambulances!" The fifth alarm was ordered. Soon the call went out: "Send every available ambulance."

Five members of the Boston Fire Department were killed: Fire Lieutenant John J. McCorkle of Engine Company 24, age fifty-three with twenty-seven years of service; Fire Lieutenant John J. Geswell of Ladder Company 26, age forty with nine years of service; Firefighter Francis L. Murphy of Engine Company 24, age forty-two with seven years of service; Firefighter James B. Sheedy of Ladder Company 4, age thirty-seven with six years of service; and Firefighter Robert J. Clougherty of Engine Company 3, and son of the chief of department, age thirty-one with four years of service.

A twenty-five-year-old "spark," or fire buff, was also killed. Andrew "Andy" Sheehan of Milton, Massachusetts, used to hang around the quarters of Engine 3 and Ladder 3, where he was very well liked. "He died doing what he loved: helping the firefighters," his grieving father said.

Assistant Chief John Clougherty and Deputy Chief Frederick Clauss directed operations after the initial collapse at the Trumbull Street Fire. Tragically, Chief Clougherty's son Robert of Engine Company 3 was killed, and Deputy Clauss was seriously injured in a second collapse.

A large funeral was held on October 5, 1964, for all five firefighters at the Cathedral of the Holy Cross on Washington Street in the South End, only a few blocks from the fire scene. A funeral was held on the same day for Andy Sheehan in Milton, Massachusetts.

The cause of the fire was never officially determined, but it was suspected that kids had been in the building earlier in the day and possibly had ignited some material that might have smoldered for hours.

A plaque adorns the outside of Florian Hall, the union hall of the International Association of Fire Fighters, Local 718, in tribute to these five fine firefighters. The name of Andrew Sheehan also appears on the plaque.

[B.N.]

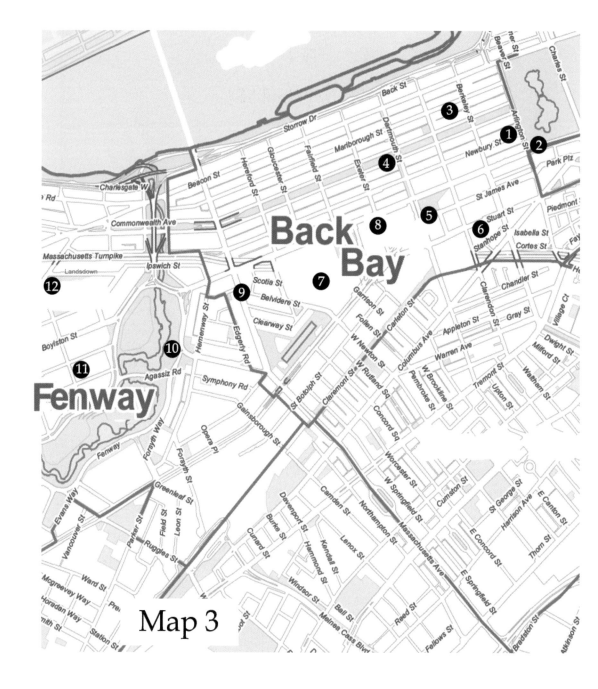

Map 3

Back Bay/Fenway

1. Arlington Street Fire

16–17 Arlington Street/0 Newbury Street,
corner of Arlington and Newbury Streets, Back Bay

On January 6, 1981, shortly after 3:00 p.m., a fire was discovered in a commercial/retail building at the corner of Newbury and Arlington Streets in the Back Bay. The building is a four-story brownstone, typical of the Back Bay, and was built in 1869. The location is at the beginning of one of Boston's most upscale shopping districts. Box 1539, Newbury and Berkeley Streets, was struck at 3:07 p.m. and Fire Alarm announced that calls were being received for "a building at Arlington and Newbury Streets." Arriving units reported heavy smoke showing.

Fire spread through an open-cage elevator shaft to the upper floors, which housed several offices, including that of the Boston School Volunteers and former Governor Francis Sargent. Governor Sargent was able to escape the building unharmed, but his extensive collection of political files was severely damaged. Six alarms were struck in order to cut the fire off from spreading in two directions, due to the fire building being located at the corner of two streets.

District 4 Chief Thomas Dunbar and Division 1 Deputy Chief Gerald Hart were the incident commanders at the scene. After the fire had been knocked down and overhauling operations had begun, a partial collapse of the upper floors occurred (an event similar to the tragic Vendome Fire of 1972). The third floor gave way, with the fourth floor crashing down on top of the firefighters. Seventh and eighth alarms were ordered. Twelve firefighters were trapped in the rubble. The body of Fire Lieutenant Paul M. Lentini of Engine Company 37 was located, trapped by a fallen beam. Beneath Lentini were several firefighters, who were trapped but alive. Searches continued for other missing firefighters, until all were accounted for except for Firefighter James M. Gibbons of Engine 37. After many hours of searching, his body was recovered at 10:30 p.m. at the bottom of the collapse area. The final toll was two deaths and ten injured firefighters, four seriously.

Engine 33's horse-drawn steam pumper responds southbound on Massachusetts Avenue, crossing Huntington Avenue in the Back Bay, circa 1910. Horticultural Hall is visible in the background. Horse-drawn apparatus in Boston began to be phased out in 1910.

The site of the fire/collapse was rebuilt in a manner consistent with the surrounding properties. An observer of the location today would be hard-pressed to detect any evidence of the fire and subsequent collapse. The site remains an active retail and commercial enterprise, occupied by a Burberry store and Bierly-Drake Interiors store. [M.G.]

2. Statue of William Ellery Channing

Public Garden, corner of Arlington and Boylston Streets, Back Bay
One of the loveliest spots in Boston is the Public Garden, which was built on a landfill in the nineteenth century. It contains the city's famed Swan Boats, among its many attractions. Its sculptures include renditions of the family of fowl from *Make Way for Ducklings* and Thomas Ball's statue of the father of our country (and volunteer fireman), George Washington. In the southeast corner, facing the Arlington Street Church, is a statue of famed preacher and philosopher William Ellery Channing, who influenced

Engine 26 operates at the corner of Arlington and Newbury Streets on a cold January 6, 1981. Two firefighters perished in a partial collapse of the upper floors at this eight-alarm fire.

The Fire Alarm Office at the Bristol Street fire headquarters was opened on May 20, 1895. Beautifully detailed with parquet flooring, it occupied the top floor. Critics denounced the location, as it was very near to combustible structures. Despite the Blacker and Shepard fire, the office would remain here until 1925.

such thinkers as Ralph Waldo Emerson. Channing may have saved many souls, but it was his son, William Francis Channing, who saved countless lives. That's because in 1845 the younger Channing, a doctor, came up with the idea of applying the newly invented telegraph to create a municipal fire alarm system. Previously, the only way to get the word out about fire was to (literally) run down the street yelling "Fire" and to ring out church bells to direct firefighters to a general location. The system was terribly inefficient, and Channing designed a system in which a municipal telegraph system could relay information to a central Fire Alarm Headquarters, which could dispatch companies to a specific location. The city council agreed to fund the idea and Channing, with the mechanical genius of engineer Moses Gerrish Farmer, built the world's first municipal alarm system in 1851. It worked beautifully. Soon alarm systems were being built around the country, many by the Gamewell Company. Channing's vision and Farmer's engineering were the first steps in creating a modern public safety communications system. The first Fire Alarm Office was in Court Square, behind city hall. It was later moved to various locations, including Old City Hall and the department headquarters on Bristol Street in the South End; in 1925, it was relocated to a building in the Fenway. In all the years, the fire alarm system has never failed once. Yet the names of William Francis Channing and Moses Farmer are little known, except to Boston firefighters.

[S.S.]

3. First Church Fire

288 Berkeley Street, 66 Marlboro Street and
29–39 Commonwealth Avenue, Back Bay
A spectacular five-alarm fire occurred on March 29, 1968, about 2:17 a.m. at the First Church of Boston. Firefighters found the church fully involved when they arrived; the fire also spread to an eleven-story office building and apartment buildings on Marlboro Street and Commonwealth Avenue. The church's prominent stained-glass window blew out due to the heat of the fire. The recently inaugurated Mayor Kevin H. White was on scene, as was Fire Commissioner William Fitzgerald. This was the sixth five-alarm fire recorded in the city of Boston during the first quarter of calendar year 1968. Although the fire occurred during the early morning hours, with five alarms sounded in record time, there was, amazingly, no loss of life. The First Church of Boston, Unitarian Universalist was rebuilt as a modern structure while preserving the original façade and outline of the large, round stained-glass window facing Berkeley Street, which had been destroyed in the fire. The buildings facing Commonwealth Avenue have all been restored and look pretty much the same today as they did in 1968.

The First Church of Boston was fully involved in fire on the arrival of the first company, Engine 10, shortly after 2:00 a.m. on March 29, 1968. Engine 2's Mack pumper is visible in the foreground.

[B.M.]

4. The Vendome Fire

Vendome Hotel and Vendome Monument,
corner of Dartmouth Street and Commonwealth Avenue, Back Bay
Saturday, June 17, 1972, was a beautiful spring day with bright sun and warm temperatures, but this day would become one that firefighters throughout the area would never forget. A fire originated in the Vendome Hotel, a six-story building that was undergoing renovation. When the first fire companies arrived, smoke was issuing from windows on the fourth and fifth floors and firefighters started their operations with hand lines over ladders.

This photo shows chief officers directing recovery efforts at the Hotel Vendome fire on June 17, 1972, after the building collapsed onto Ladder 15. Nine firefighters died in the collapse. On the left, note the height of the rubble within the building.

Box 1571 was transmitted at 2:35 p.m. District Fire Chief William Doherty of District 4 reported a working fire at 2:44 p.m. and then ordered a second alarm at 2:46. Deputy Fire Chief John J. O'Mara of Division 1 ordered the third alarm at 3:02 p.m., and the fourth alarm at 3:06 p.m.

Conditions forced firefighters from inside the building. Ladder Company 15 took a position in the rear alley and raised the aerial to the upper floors. Ladder Company 17 was on Dartmouth Street, along with Aerial Tower 2, and Aerial Tower 1 was on Commonwealth Avenue along with Ladder Company 4.

As the fire was brought under control, firefighters were again sent inside to mop up the pockets of fire. Suddenly, at 5:28 p.m., the entire southwest corner of the building collapsed, trapping roughly a dozen firefighters in the rubble.

The Fire Alarm Office was notified of the collapse and the Rescue Pumper was requested, along with the Cambridge Fire Department Rescue Company. Chief of

This impressive monument on the Commonwealth Avenue Mall memorializes the nine firefighters lost at the Vendome Hotel fire. It was dedicated on June 17, 1997, the twenty-fifth anniversary of the fire.

Department George H. Paul was on vacation, but he was notified and responded to the fire scene, as did Fire Commissioner James Kelly.

Once the word got out that firefighters were trapped in the collapse, many off-duty firefighters rushed to the scene and assisted with the digging. For hours, the men worked frantically to save their fellow firefighters.

Nine members of the Boston Fire Department were killed: Fire Lieutenant Thomas J. Carroll, Engine Company 32, age fifty-two, with twenty-seven years of service; Fire Lieutenant John E. Hanbury, Ladder Company 13, age fifty-two, with twenty-three years of service; Firefighter Richard B. Magee, Engine Company 33, age thirty-nine, with four years of service; Firefighter Thomas W. Beckwith, Engine Company 32, age thirty-five, with six years of service; Firefighter Joseph P. Saniuk, Ladder Company 13, age forty-seven, with twenty-four years of service; Firefighter Charles E. Dolan, Ladder Company 13, age forty-seven, with twenty-five years of service; Firefighter John E. Jameson, Engine Company 22, age fifty-two, with twenty-one years of service; Firefighter Paul J. Murphy, Engine Company 32, age thirty-six, with five years of service; and Firefighter Joseph F. Boucher, Engine Company 22, age twenty-eight, with nineteen months of service. By daylight Sunday morning, eight women were widowed and twenty-eight children had lost their fathers.

This view of the Vendome is from the corner of Dartmouth and Newbury Streets at the height of the fire. This tragic fire and the subsequent building collapse that killed nine Boston firefighters remains vivid in the minds of the Boston Fire Department community.

It took close to twelve hours to find and remove all of the men in a heartbreaking and backbreaking night for the firefighters. The most seriously injured was Fire Lieutenant James McCabe of Engine Company 33.

A large funeral was held for all of the firefighters at the Cathedral of the Holy Cross on Washington Street in the South End on June 22, 1972. Wind and rain from the remnants of Hurricane Agnes buffeted the more than ten thousand firefighters from all across the country and Canada who came to pay their respects.

After years of work by Deputy Fire Chief Paul Christian and District Fire Chief Gerard Molito, the city agreed to install a monument dedicated to the Vendome firefighters on the Commonwealth Avenue Mall, within sight of the rebuilt hotel, which is now a condominium complex. The Vendome Memorial was dedicated on June 17, 1997, twenty-five years to the day after the fire, in a ceremony attended by thousands of firefighters and their families. The memorial was designed by sculptor Ted Clausen and landscape architect Peter White, working in collaboration as Third Floor Studio. It was paid for by individual donations, corporate sponsors and the sale of many T-shirts by both active and retired firefighters. The Browne Fund of the city also made a substantial contribution to the memorial. It features a bronze cast of a firefighter helmet and a turnout coat draped over a granite arc decorated with quotes and statements that refer to the fire. Unfortunately, Chief Molito died before the monument was finished.

As you contemplate the memorial, located across the street from the Vendome Hotel on Commonwealth Mall, you cannot help but think of the nine firefighters who went to work that day but never returned.

[B.N.]

5. Two Fires: Copley Plaza Hotel Fire and Sheraton Boston Hotel Fire

205 Dartmouth Street, Copley Square; and 39 Dalton Street, Back Bay
During the early morning hours of March 29, 1979, the sixteenth anniversary of the Sherry Biltmore fire, there were two nearly simultaneous hotel fires in the city, within blocks of each other.

The first fire was in the seven-story Copley Plaza Hotel at 205 Dartmouth Street in Copley Square. The second fire was in the Sheraton Boston Hotel at 39 Dalton Street. At the time, the Copley Plaza Hotel was occupied by 430 persons, all of whom were evacuated, with at least 50 people taken over ladders by firefighters. About 30 persons were injured; 1 of the victims later died. The Sheraton Boston was occupied by about 1,400 persons, 24 of whom were injured. All evacuations at this fire were accomplished using interior stairs.

Just after 1:00 a.m., Engine Company 33, Ladder Company 15 and District 4 were dispatched to a fire at the Copley Plaza Hotel. The first companies were directed to several separate fires in the basement, which were handled very quickly. Another fire was discovered on the fourth floor and was making headway, causing patrons to be trapped. When District Fire Chief Thomas Dunbar was informed of the major fire on the upper floors, he ordered the second alarm; eventually five alarms would be ordered.

Ladder trucks were positioned and rescues were made from the upper floors. Ladder Company 13 was positioned at Stuart Street and Trinity Place. The aerial ladder was raised, and several people were removed from the upper floors. Smoke was blowing across the front of the hotel, which made the job of positioning the aerial ladders difficult. While people were being taken over ladders, other firefighters were working in the interior on the stairwells, assisting occupants down through the dense smoke.

The fire was charging down the hallway like a freight train. It blew out several windows at the corner of the hotel facing Copley Square. Members of Engine Company 10 took a line of hose over the aerial ladder of Ladder Company 15 and hit the heavy fire from the outside, knocking it down.

Firefighters advance over the aerial ladder of Ladder 30 into the fourth floor of the Copley Plaza Hotel at St. James Avenue and Dartmouth Street. Many rescues were made via aerial ladders at this fire, the first of two major hotel fires in the early morning hours of March 29, 1979.

The building had a courtyard in the center that was several stories above the ground. Several people used sheets that were tied together to slide down to the lower roof, making their way to the other side of the hotel, and thus escaped safely. Several members of the department also made their way onto this roof and exited the area when the flames turned down the hallway.

While all of the fire companies were still working at this fire, another alarm was transmitted for another hotel, the Sheraton Boston at 39 Dalton Street, not far away. Arriving firefighters found a fire burning in the Falstaff Room of the hotel, with a heavy smoke condition. District Fire Chief Vincent Bolger reported a working fire at 2:36 a.m. Several separate fires were apparently set in this hotel, but these were controlled pretty quickly. The main fire was rapidly brought under control, but the upper floors of the hotel were filled with heavy smoke. Three alarms were transmitted for this fire and several companies worked at both fires.

The Boston firefighters received much praise from the news media for their performance at both fires. A suspect was arrested and charged with setting the fires in both hotels. The motive was revenge, as he had worked at both hotels and been fired from both. Both hotels underwent extensive renovations and continue in use today as desirable lodging for those visiting Boston.

[B.N.]

6. Red Coach Grill Fire

41–45 Stanhope Street, Back Bay

The Red Coach Grill was located at 41–45 Stanhope Street, right behind Boston Police Headquarters. The morning of July 1, 1955, was a very hot and humid day, with temperatures expected to reach close to one hundred degrees.

Just before the Boston Fire Department shift change that morning, about 7:42 a.m., Box 1547 was transmitted for a fire in the Red Coach Grill. Smoke was pouring from the basement. Ladder Company 13 arrived at the front of the building with a 1950 American LaFrance eighty-five-foot wooden aerial and the "stick," or aerial ladder, was thrown to the roof. As the smoke grew heavy, a second alarm was ordered. Ventilation was difficult and holes were cut in the roof, but the smoke just did not lift. The restaurant had a limited number of windows available to ventilate the heavy smoke and heat. The engine companies were having trouble getting into the basement, where the fire was located. At 8:04 a.m., a third alarm was ordered, as the weather and heavy smoke were taking their toll on the firefighters.

As firefighters collapsed from the heat and smoke, extra companies were called. The narrow stairways to the basement area made fighting the fire difficult. The department's filter masks—known as All Service Masks—did not give the firefighters the protection afforded with today's masks. Firefighters were forced to wear the cumbersome Chemox masks, which were reserved for fires in ships, subways or subcellar fires. Two extra engine companies were called, and then four extra engine squads were called. The rescue company was also called for use of its exhaust fans and other special equipment. Firefighters working on the roof ventilating were near exhaustion and collapse from the high heat and humidity.

Fire Commissioner Francis X. Cotter arrived and announced that anyone who would stay on duty after 8:00 a.m. would be put on overtime. This was unheard of in those days. Firefighters were normally given time off; but if you went to the hospital, your overtime was stopped as soon as you got into the ambulance.

The smoke, along with the heat and humidity, made this one of the city's nastiest fires. Twenty-five firefighters were treated at Boston City Hospital. The all out was not sounded until 5:10 p.m., and the damage was listed at $150,000, a huge amount for the year 1955.

[B.N.]

7. Prudential Center

800 Boylston Street, Back Bay

The Prudential Center—with its observation deck above and shops below—has been a prime tourist attraction in Boston since it was built in 1965. But "the Pru" also has a place in Boston's fire history. At 750 feet and fifty-two stories, the Prudential was Boston's tallest building until 1976, when the John Hancock Tower was built in Copley Square.

During a three-alarm fire on July 1, 1955, a medical aid station was set up across the street from Red Coach Grill. The fire was fought under extreme heat, high humidity and heavy smoke conditions.

About 5:00 p.m. on January 2, 1986, a security guard discovered that a fire alarm had activated on the fourteenth floor. Responding firefighters could see flames on the fourteenth floor, and a second alarm was ordered. Eventually eight alarms were struck. About fifteen hundred people worked in the building, which posed the risk of a huge tragedy. When tenants, smelling smoke, began to evacuate, they found that stairwells were filled with smoke. The smoke was not only rising to the upper floors, but was also banking down, even to the first floor. Firefighters shared their oxygen with hundreds of panicked office workers who were trying to get down the stairwells. Fortunately, no one was killed, but about twenty-four people were injured. The fire apparently started in construction materials in an unoccupied floor that was being renovated. The Pru was retrofitted with a sprinkler system, and the city passed an ordinance requiring that any building higher than seventy feet in Boston needed to install a sprinkler system.

This wasn't Boston's only problem with high-rise buildings. The nearby sixty-two-story John Hancock Tower also had its public safety problems due to a design flaw. After the I.M. Pei–designed structure was built in 1976, windows would suddenly pop out of the upper floors and rain glass below. All of the building's thousands of windowpanes were removed and replaced with fire-resistant plywood until new windows could be re-engineered and installed. For a considerable time, Boston had the tallest plywood skyscraper in the world. The problem has been since corrected.

[P.C., S.S.]

A firefighter activates a self-generating Chemox oxygen mask prior to entering the basement of Red Coach Grill on Stanhope Street. Twenty-five fighters were sent to Boston City Hospital for treatment during this July 1, 1955 fire.

8. Lenox Hotel Fire

50 Exeter Street, corner of Boylston Street, Back Bay

At Boylston and Exeter Streets in the Back Bay stands the eleven-story Lenox Hotel. Built in 1900 and still in use today, it is one of the oldest hotels operating in Boston. It was the scene of a spectacular four-alarm fire early on the cold Saturday morning of February 10, 1917, in which many spectacular rescues were performed in twelve-degree temperatures by members of the Boston Fire Department. Outstanding ladder work was performed by firefighters assigned to Ladder Companies 13 and 15.

The fire started in a guest room at the rear of the third floor at about 4:00 a.m., and the occupant, in a hurry to escape, left his door open, which caused the fire to spread rapidly to the corridor and upper floors, trapping many guests in their rooms. More than 250 people were staying in the hotel that night.

The first alarm from Box 1573 was sounded at 4:59 a.m., and Engine Company 33 and Ladder Company 15, arriving on the scene from quarters at Boylston and Hereford Streets, found heavy fire and smoke showing from every side of the building. People were visible in windows at every level above the third floor, waiting for rescue. Ladder 15 raised the company's eighty-five-foot aerial ladder on the Boylston Street side of the building and began to make rescues, moving the ladder from window to window and pulling people from the window ledges. Fire Captain Charles Donahue of Ladder

15 took a pompier ladder and lifeline and went to the top of the aerial ladder, which extended only to the seventh floor. He then used the pompier ladder and went up floor by floor on the exterior and made additional rescues at the ninth, tenth and eleventh floors, bringing numerous persons to the street by this means.

Ladder Company 13, also arriving on the first alarm with Engine Company 22 from their quarters on Warren Avenue, placed the aerial ladder on the Exeter Street side. They also used pompier ladders for rescues from the upper floors on that side of the building. Many more were rescued by members of Ladder Company 17 while engine companies stretched lines to the interior to fight the fire.

As soon as it was determined that all residents had been removed to safety, heavy-stream appliances were put into operation. Water from deck guns and water towers was used to darken down the heavy fire that was showing from windows on the upper floors. One resident died of a heart attack after returning to his room, but all others were safe, and either made their own way to the street or were rescued by firefighters. Additional alarms were struck on Box 1573 at 5:09 a.m., 5:17 a.m. and 5:19 a.m., bringing additional companies to the fire.

Special mention was made in general orders, published by the fire department several days later, of rescues made by individual firemen at the fire. Included was the story of Ladderman John. J. Kennedy of Ladder 13, as follows:

> Ladderman Kennedy finding that he was unable to reach entrapped guests from the top of the 85 foot aerial ladder and unable to use a pompier ladder to ascend higher, owing to the projection of an ornamental cornice, at personal risk, climbed over and above the cornice, to a window on the 8th floor, from whence he lowered Mr. and Mrs. Finestone with a lifeline to members of his company at the top of the 85-foot ladder. For this remarkable manifestation of courage, Ladderman John J. Kennedy is placed on the Roll of Merit.

[J.T.]

9. Sherry Biltmore Hotel Fire

150 Massachusetts Avenue, near Belvidere Street, Back Bay

One of the most spectacular jobs of raising ladders was done by the members of the Boston Fire Department who were working the shift during the morning of Friday, March 29, 1963, when a fire broke out in the Sherry Biltmore Hotel at 150 Massachusetts Avenue, on the corner of Belvidere Street, in the Back Bay. The eight-story building was built in the early 1900s as an apartment hotel; it was remodeled and reopened in 1955 as the Sherry Biltmore Hotel. The bottom three floors contained stores, lounges, ballrooms, lobby and a kitchen, as well as other functional and service areas. The upper five floors contained guest rooms. The hallways had been covered with inch-thick plywood and had a suspended tile ceiling.

The building had a courtyard in the middle. The center of the ring was once an open court with an archway at the ground level that led directly to the street. During reconstruction, this arch was bricked up and a health club was built into the court.

The fire was reported to have started in an unoccupied room, room 655 of the sixth floor, and went unnoticed for several minutes. The occupant of room 644 reported a smell of smoke; she opened the door and could see a light haze at the ceiling, so she called the switchboard operator. The night bellboy was sent to the sixth floor to investigate. He found so much smoke in the corridor by the door to room 655 that he could not see the room number. He reported this to the manager at the front desk, who was busy telling guests on the upper floors "not to worry." About 3:55 a.m., a guest who had been awakened by the smoke pulled the break glass station and fled toward the exit near room 630. But this was a local alarm to alert guests, and it did not transmit to the Boston Fire Department.

The bellboy raced back to the sixth floor, this time with a fire extinguisher, but the heat was so intense that he couldn't get close to the fire. The hotel manager lived in room 628. When she saw the fire, she screamed "Run, run!" to the bellboy. But she couldn't get out, and she sought refuge in room 644.

The hotel was nearly fully occupied at the time of the fire, and the thick smoke was filtering through the halls and rooms. People were trapped in their rooms and went to the only place they thought they could get out: the windows. A party was still going strong in suites 603 through 611 for a departing member of the cast of *The Sound of Music*. Suddenly, one of the girls smelled smoke and opened the door of room 603 to find heavy smoke in the corridor outside and a tongue of flame near the top of the door. She slammed the door and ran to the door for room 611; it, too, was clogged with heavy smoke. The now-panicked partygoers began to smash out windows; one person found a small window in the bathroom that was partially over a roof outside. Crawling out head first, one at a time, the group managed—with great difficulty—to get out and onto another roof, where they were saved from certain death.

Occupants trying to get out and firefighters trying to reach people inside had to cope with window air conditioning units that blocked their way. Several occupants kicked out the units and were able to get to lower roofs and then to safety. Others were not so fortunate.

Of the twenty-six persons known to have rooms overlooking the sixth-floor inner court, two died in their rooms. Three were rescued over a makeshift ladder. Two boys, in an area that the ladder could not reach, slid down poles outside their windows to safety. One man tied his bedding together and went out the window, but the sheet ripped and he dropped about twelve feet. Thankfully, he escaped injury.

About 4:00 a.m., a Boston Police officer saw persons leaning out of the windows and heavy smoke coming from windows on the sixth floor. He radioed the Boston Police dispatcher to "send all ambulances and notify the Fire Department that the Sherry Biltmore is on fire." The delay in notification may have cost lives.

Fire Captain Leo Wisentaner of Ladder Company 15 arrived from quarters only a few blocks away on Boylston Street and reported a working fire at 4:02 a.m. He also ordered another truck to the fire.

This hotel at Massachusetts Avenue and Belvidere Street was the scene of one of the most extensive "laddering jobs" in the history of the Boston Fire Department. Ten ladder companies responded to a fire at the Sherry Biltmore Hotel on March 29, 1963, and although many rescues were made, four persons died.

District 4 Fire Chief John McCarthy arrived. He took one look at the blaze and ordered a second and third alarm. Division 1 Deputy Fire Chief James "Fuzzy" Flanagan of Division 1 ordered the fourth alarm at 4:11 a.m. and the fifth two minutes later. Six additional ladder companies were requested; ladders 3, 30, 20, 18, 23 and 1 responded.

Firefighters feared that people would start jumping from the windows. District Fire Chief McCarthy ordered Ladder Company 13 to stop at Massachusetts Avenue and Belvidere Street, where firefighters raised the aerial and removed people from the upper floors. Engine Company 22 raised a forty-foot ladder on St. Cecilia Street and firefighter Albion Burke removed a woman.

By then, fire had engulfed the sixth-floor corridor and was sending flames into the guest rooms. Compounding the critical situation, fire was appearing at various exterior windows and was beginning to overlap into the floors above. Additional engine companies and a water tower were called in.

Thanks to the efforts of the firefighters who gave their all at this fire, only four people died. At least twenty-seven people were injured and transported to Boston City Hospital. Several ladder companies used every ladder carried on the apparatus; additionally, ladders were carried up the side of the building and set up on decks above the first floor. An estimated one hundred people were rescued, many over ladders, while some were removed by the interior stairs. Others made it out on their own. The building is now used by the Berklee College of Music.

[B.N.]

10. Boston Fire Alarm Headquarters

59 The Fenway, Fenway

The current Boston Fire Alarm Headquarters in the Fenway neighborhood of Boston was the culmination of a search that stretches back almost to the origin of the fire alarm system itself. In the early 1860s, when the new city hall was being constructed at 45 School Street, space in the upper story was allocated for a new and improved Fire Alarm Office. It was opened on December 26, 1865. A report by the National Board of Fire Underwriters in 1890 noted that the office was located in a congested part of the city in the city hall dome, constructed almost entirely of wood. The fire commissioner and fire alarm superintendent were aware of these conditions, but had been unable to find a new site. In the early 1890s, a new fire headquarters was planned at 60 Bristol Street in the South End. On May 20, 1895, the new headquarters opened, with the Fire Alarm Office in the top floor of this four-story fire-resistive building.

The National Board of Fire Underwriters, however, remained concerned. Their November 1905 report noted that while the building was of excellent construction, it was located across a forty-foot-wide street from a woodworking business. Within a few years, their fears were realized when on August 9, 1910, a serious fire destroyed the Blacker and Shepard lumberyard on Albany Street. The fire spread to and destroyed

The current Fire Alarm
Office was opened on
December 27, 1925, at
59 The Fenway, opposite
Westland Avenue in the Back
Bay/Fenway area. The new
structure was built expressly
for fire alarm operations
and was located more than
250 feet from the nearest
structure to ensure its safety.

the fire department's repair shop at the northwest corner of Albany and Bristol Streets, threatening fire headquarters itself. Fortunately, fire headquarters escaped damage, but this incident triggered pressure for the removal of the Fire Alarm Office to a safer, isolated location.

Various attempts were made to secure a new site; the current Fenway location was not selected until 1923. Since the proposed site was on park department land, a special act of the legislature was required to permit construction, which began with groundbreaking ceremonies on April 1, 1924. In accordance with the standard then prevailing, the new building was located 250 feet away from any other structures in the area and was designed with a view toward beauty, utility and compatibility with its surroundings. Its dimensions were sufficiently large to accommodate expansion for many years to come. Construction had progressed sufficiently in 1925 to permit the Gamewell Company to begin installation of fire alarm equipment in the building on April 22 and for the telephone company to install cables on May 19. All fire alarm circuits had been connected to the new office by September 28, 1925.

At 8:00 a.m. on December 27, 1925, the new Fire Alarm Headquarters was officially put in service by Mayor James M. Curley, Chief of Department Daniel F. Sennott, Superintendent of Fire Alarm George Fickett and Fire Commissioner Theodore A. Glynn, and the Bristol Street office was permanently discontinued. The first—and ceremonial—alarm transmitted from the new Fenway office was at 8:02 a.m. from Box 2328 on Westland Avenue. This box was sounded by the aide to the fire commissioner and was ordered transmitted to the department.

Proudly and somewhat defiantly, the following words are cut into the limestone frieze facing Westland Avenue: "ERECTED BY THE CITIZENS OF BOSTON TO FORTIFY AND EXTEND THE PRINCIPLE OF ORGANIZED RESISTANCE TO THE SCOURGE OF FIRE, CONSECRATED AND DEDICATED TO THE SERVICE THROUGH WHICH THIS PRINCIPLE IS SO NOBLY PERPETUATED."

Today, the Fire Alarm Office continues to do the job it was designed to do, albeit with many internal modifications, utilizing the newest and most effective fire communication technologies available. The motto of the Fire Alarm Office, "The eye that never closes," is as true today as it was in April 1852, when the system became operational. Recently the security of the office has been enhanced due to terrorism concerns; as a result, access to the building is greatly restricted. However, if one could gain permission for a visit, one could view a beautiful restored display of antique and vintage fire alarm equipment that represents the various innovations and advances that have been added to the system over the years.

[T.G.]

11. The Peterborough Street Fire

50 Peterborough Street, corner of Jersey Street, Fenway
On March 31, 1971, at 12:33 a.m., Box 2341 alerted firefighters of a fire at a five-story apartment building on Peterborough Street. Captain Leo Wisentaner of Ladder 15

An apartment house fire at 50 Peterborough Street in the Fenway area on March 31, 1971, resulted in five alarms being transmitted. Many rescues were made; however, eight persons perished.

reported a working fire while en route to 50 Peterborough Street. He found occupants in windows, on balcony fire escapes and sitting on window ledges of all floors, particularly on the Jersey Street side, awaiting rescue.

Wisentaner could see fire through the roof in the center of the building, as well as in all hallway windows in the rear, in the fourth-floor rear left and the fourth-floor right side. Two first-floor apartments and the first-floor hallway were fully involved.

Third, fourth and fifth alarms were ordered by District Chief Kennealey, Deputy Chief Magoon and Chief of Department George H. Paul. Exemplary ladder work of the Boston Fire Department saved countless lives. Despite heroic efforts, eight lives were lost. A new apartment building stands on the site today.
[B.M.]

12. Fenway Park Fire

Fenway Park, Yawkey Way, Fenway
Everyone knows that the Red Sox play baseball at Fenway Park, but did you know about the huge fire there in 1934?

The weather was cold, with some snow falling, on Friday, January 5, 1934. Fenway Park was undergoing a major renovation that would cost over $250,000, a substantial

On January 5, 1934, Fenway Park was undergoing renovation. The left field bleachers became the scene of a fire that spread rapidly. This view of Lansdowne Street looking toward Brookline Avenue shows how the fire jumped the street, igniting buildings on the opposite side.

amount in the 1930s. More than 150 workers were on the job site when a salamander heater that was being used to dry the fresh cement overturned. A canvas covering caught fire, and the fire spread rapidly. The workers tried in vain to put it out, but the blaze quickly reached the new left field bleachers in the area known as Duffy's Cliff. At 1:04 p.m., Box 2344 was pulled, sending firefighters racing to Fenway Park. The first fire companies to arrive found heavy fire in the area of the new bleachers, and eventually a fifth alarm was ordered at 1:24 p.m. by Chief of Department Henry A. Fox.

Companies had to run lines through the park from Ipswich Street in an attempt to stop the flames. But the fire jumped Lansdowne Street and took possession of several buildings, including the Seiberling Rubber Co., the old Cotton Club and the Oldsmobile & Pontiac Motor Works located at 42 Lansdowne Street. Companies worked the deck guns as they advanced up Lansdowne Street. (In 1934, every engine company in Boston had both a hose wagon and a pump.) Eddie Collins, general manager of the Red Sox, was having his lunch when he was notified of the fire, and he rushed to the park. Team owner Tom Yawkey was on vacation in North Carolina when he was notified, and he headed for Boston immediately. When he arrived in Boston, he announced, "Opening day will not be delayed."

Fire Commissioner McLaughlin, who had been reappointed to that position on the previous day, left his "welcome back lunch" to respond to the fire. Only after a long and

Fenway Park, *left*, shows fire and heavy smoke in this view of Lansdowne Street on January 5, 1934. Opening day in April, however, went off without a hitch.

hard battle was the Fenway fire brought under control. The damage exceeded $220,000, an astronomical loss in 1934. Two firefighters received minor injuries while battling the inferno.

Repairs to the stadium were completed on time, and as Tom Yawkey promised, opening day was not delayed.

[B.N.]

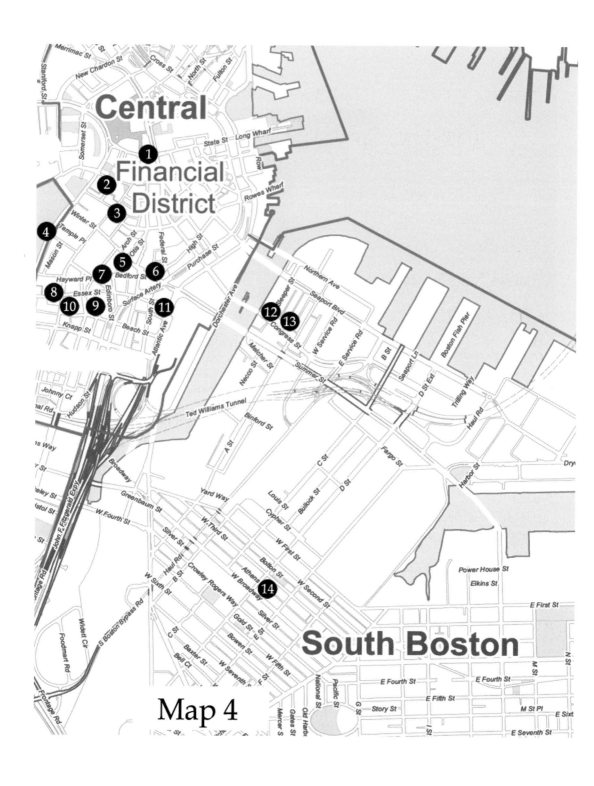
Map 4

Map 4

Downtown Crossing/Financial District

1. The Old State House Museum

206 Washington Street at State Street, Downtown/Financial District

Fire historian Paul Ditzel once said Boston was "built to burn." That's because the town—crowded onto a peninsula—had rows of tightly packed homes that meant fire could spread easily. Early Boston settlers suffered numerous devastating fires, including huge conflagrations in 1653 and 1676. In October of 1711 on a street then called Cornhill Row (roughly where Washington Street is today), a fire started in a dilapidated hut. Spreading quickly, the blaze devastated one hundred buildings, including the fifty-year-old Town House, the seat of the government of the Massachusetts Bay Colony. A new government seat was built on the site in 1713; it continued to house the commonwealth government until the 1790s, when a new statehouse was built on Beacon Hill.

The Old State House sustained serious damage in a fire on November 21, 1832. An engraved illustration of the fire was used on Boston Fire Department certificates for many years.

Just outside the Old State House is the site of the Boston Massacre, where a handful of British soldiers fired into a crowd on March 5, 1770. The Old State House now houses a Boston history museum operated by the Bostonian Society, which has fire-related materials in its extensive archives and collections.

[S.S.]

2. Old City Hall

45 School Street, Downtown/Financial District

Old Boston City Hall is a handsome Second Empire–style building on School Street in Boston's downtown. Built in 1865, it was the second location of the Boston Fire Alarm Office. (The world's first electric fire alarm system went into service on April 2, 1852, in

An engraved illustration of an 1832 fire at the Old State House was used on Boston Fire Department certificates for many years.

the City Building at Court Square and Williams Court—directly to the rear of the lot upon which the city hall eventually would be built.) The Old City Hall has a prominent cupola located above the center of the front of the building. It was within this cupola that the fire alarm equipment was installed, with cables linking it to the fire alarm signal boxes, public bells and engine houses throughout the city. The location at the time was an architectural feature that gave the operators a decent 360-degree view of the downtown area, as well as ample natural light. Thus, in the event of a system failure, fires might also be spotted from the cupola.

Aside from being the second home to Boston's historic first fire alarm system, city hall was also the scene of many other events and political maneuvering that affected the fire department over the years. For example, after John S. Damrell was appointed chief engineer of the Boston Fire Department in 1866, he repeatedly petitioned the Board of Aldermen for improvements to the city's fire protection infrastructure. Among the requests were pleas for larger water mains, improved hydrants, additional fire engines and more fire alarm boxes. In addition, Damrell sought authority to inspect buildings and plans for new buildings. But city officials responded to the requests by accusing Damrell of empire building. His vindication came during and after the Great

These are portraits of several key city officials at the time of the Great Fire in 1872. *Clockwise from top*: Mayor William Gaston, Police Chief Edward H. Savage and Chief Engineer John S. Damrell.

Fire of November 9–10, 1872, when a devastating fire made many Bostonians realize that Damrell was a man ahead of his times. Damrell went on to be the city's first building inspector and developed the city's inspection department into a well-regarded institution.

Old City Hall, as it is now called, was closed in 1969 when the new city hall was opened in Government Center, the anchor of Boston's urban renewal efforts of the '50s and '60s. Old City Hall has housed restaurants and is a prestigious office address in downtown Boston.

[T.G.]

3. Old South Meeting House

310 Washington Street, Downtown/Financial District

The Old South Meeting House is one of the most significant historical buildings in Boston. Built in 1729 as a Puritan church, the structure has endured wars, revolutions, social change and fire. In colonial times, its congregation included the African American slave and poet Phillis Wheatley and American Revolutionaries Samuel Adams and Benjamin Franklin, whose many accomplishments include the formation of a mutual aid fire society in Philadelphia. In 1773, a group of Americans angry over a controversial tea tax imposed by the British gathered to protest at the Old South, and that night the Boston Tea Party was staged. The Redcoats later took revenge by gutting the interior and setting up a stable inside.

But the greatest threat to the Old South came on November 9 and 10, 1872, when the Great Boston Fire started just a few blocks away. Firefighters worked desperately during the next two days to save the building. Many volunteers heroically scaled the roof, where they tried to stamp out sparks coming from other buildings. Despite their efforts, the building seemed doomed. When the steeple clock chimed at 6:00 a.m. on Sunday, November 10, an onlooker exclaimed, "Dear old church, I'm afraid we shall never hear that bell again." But steam fire engines arriving from New Hampshire with fresh men were able to get more lines of water on the roof of the church. According to later reports, the engines included the Kearsarge Company 3 from Portsmouth, New Hampshire. With the help of the reinforcements, the building was saved, even though the area around it was devastated.

Today the Old South continues to host events and lectures on Boston's history, particularly issues of free speech. Exhibits inside also explore aspects of New England's colorful past. Nearly three hundred years later, the Old South remains an active community meeting place for the exchange of ideas. All that is possible because of the efforts of volunteers and firefighters during one of the city's most destructive fires.

[S.S.]

The C. Crawford Hollidge department store at 141 Tremont Street, downtown, was gutted by a major fire on February 18, 1967. Five alarms in six minutes were struck for this fire, which threatened to spread to nearby stores and buildings.

4. C. Crawford Hollidge Fire

141 Tremont Street, Downtown/Financial District
Light snow was blowing across Boston as the frigid, windy morning of February 18, 1967, broke. The night crews at Boston firehouses were awaiting their reliefs for the day tour. Many of the fire lieutenants were anxious to get to the examination center for the fire captain's test, which was scheduled for 10:00 a.m.

Shortly after 7:00 a.m., the circuits at Boston Automatic Fire Alarm Company on Batterymarch Street came alive. A "trouble signal" was received from the automatic alarm system at 141 Tremont Street, the C. Crawford Hollidge department store, a fashionable women's clothing store. As the operator verified the signal and initiated a phone notification to the building, Boston Fire Alarm began striking Box 1441; simultaneously, an automatic fire alarm signal was received at Boston Automatic for the seven-story C. Crawford Hollidge building. At 7:10 a.m., Fire Alarm announced that calls were being received for a building at Tremont Street and Temple Place.

Responding from the Cambridge Street firehouse was District 3 Chief William Doherty, with Engine 4 and Ladder 24; from the Broadway firehouse Engine 26 and Ladder 17 responded; and filling out the first alarm was Deputy Fire Chief George Thompson, along with Rescue 1 and Engine 25 from the Fort Hill Square firehouse.

Fire companies winding their way down Tremont Street spotted a huge "loom up" in the gray morning sky. As District 3 Chief Doherty arrived on Tremont Street, he encountered three floors of heavy fire showing from the C. Crawford Hollidge department store. With the fire rapidly extending upward, the building was doomed. On the corner of narrow Temple Place, the inferno created serious exposure problems to nearby properties. A second alarm was struck at 7:14 a.m., and the first arriving engines were ordered to begin operating heavy-stream appliances. Deck guns were put into operation in an effort to knock down the heavy radiant heat and prevent the fire from extending to other buildings. Deputy Chief Thompson, arriving as the second alarm was striking, ordered the third alarm at 7:15 a.m. and the fourth alarm at 7:16 a.m. All efforts were put into containing the fire to the C. Crawford Hollidge building. Window frames across Temple Place at the ten-story R.H. Stearns Co. department store were igniting, and buildings adjacent to C. Crawford Hollidge on Temple Place were threatened. At 7:17 a.m., Chief Thompson ordered the fifth alarm on Box 1441. Companies were ordered into Temple Place to operate lines in the rear of the building, a tactic that proved successful, as there was no extension of the fire toward Washington Street.

Chief of Department John E. Clougherty responded on the fourth alarm and took command. Two water towers, which were kept in reserve, were ordered to the fire. Hose lines were connected to the sprinkler system of the severely exposed Stearns building. Hand lines were stretched up into the Stearns building, and fire was contained to the window areas. Calls were placed for eleven additional engine companies, including those from Cambridge, Chelsea and Newton.

Fighting high winds, bitter cold, choking smoke and searing flames, the firefighters gradually got the upper hand, and by 11:00 a.m. the blaze was subdued. The once

A self-propelled steam engine continues to pump away at the Great Boston Fire of 1872.

proud department store stood as an ice-encrusted smoldering ruin. The intersection of Tremont Street and Temple Place was clogged with ice-covered hose lines and frozen ladders. Fire companies remained on scene throughout the afternoon and into the night, and the next morning a crane was brought in to dismantle the dangerous wreckage, which had suffered a partial collapse. The C. Crawford Hollidge was gone, but R.H. Stearns and other adjoining buildings were saved.

Today, a modern office building occupies the C. Crawford Hollidge site, but the Stearns building still stands as a testimony to the heroic struggle made that bitter cold morning four decades ago by Boston firefighters.

The captain's examination was postponed to a later date.

[P.C.]

5. The Great Boston Fire of 1872

83–85 Summer Street, near Kingston Street, Downtown Crossing

The Downtown Crossing area is one of the busiest sections of Boston, filled with department stores, shops, cafés and offices. But if you had stood at the corner of Summer and Kingston Streets in early November of 1872, you would have thought that an atomic bomb had been dropped, with damage stretching as far as the eye could see. While many people know of the terrible Chicago Fire of 1871, fewer know that a horrific conflagration swept through Boston just a year later, leaving a huge chunk of the city in ruins. The Great Boston Fire led to many changes in the way the city fought fire, and the actions of one man, Chief Engineer John Stanhope Damrell, helped to inspire changes in building codes around the nation.

The fire began on the warm fall evening of Saturday, November 9, 1872. The Boston Fire Department was already on heightened alert. A severe outbreak of horse distemper had felled nearly all the steeds in the city; this meant fire engines would have to be pulled by men, not by horses, and fire officials had drawn up elaborate plans to cover the city in the event of a fire. Sometime after 7:00 p.m., a fire was discovered in a four-story granite building at the corner of Kingston and Summer Streets. The building housed a wholesale dry goods business, so it was filled with flammable material. The structure was topped by a popular mansard roof, which is particularly susceptible to fire. The cause of the fire was never officially determined (it probably came from a spark in the boiler), but it soon spread throughout the building. About 7:10 p.m., Boston Police Officer John M. Page saw the flames; he ran to Box 52, unlocked the box and cranked the alarm. It was recorded at 7:24 p.m. A second alarm was struck at 7:29 p.m. A general alarm to call in every fire engine in the city was eventually struck at 7:45 p.m.

John Damrell was the department's chief engineer, the equivalent of today's chief of department. He lived on Boston's Beacon Hill and heard the public bells ringing out 52. He grabbed his helmet and ran to the scene. The fire had already begun to spread. The first companies on the scene were Engine Companies 4 and 7 and Hose Company 2. They attempted to hold the flames to the corner, but the fire jumped the street. By 8:00

The ruins of the Great Boston Fire continued to burn for many days. Here, a steam fire engine continues to pump water. A supply of coal has been dumped to the rear of the steamer for use in keeping the boiler hot and the water flowing.

p.m., all of Boston's twenty-one engine companies were at the fire. Companies also came from Charlestown and Cambridge and, in response to frantic telegrams, fire companies from as far as Connecticut and New Hampshire started the journey to Boston. Despite their best efforts, the blaze—which was by then a virtual hurricane of flame and smoke—pressed toward Washington Street and into the heart of Boston's financial district. Firefighters were hampered by poor water pressure from the aging water mains in the area. Damrell had been pestering Boston's city council about improving water in the area, but he was told not to "magnify the needs of your department." The fire got so hot that granite in the many buildings exploded, raining masonry, embers and other debris upon the valiant firefighters. The fire became so intense that it created its own winds, defying the prevailing breezes.

Despite the danger, huge numbers of spectators filled the streets. Many merchants, seeing the spread of the flames, were trying to move supplies out of their own businesses. Some who saw flames creeping inexorably closer began giving away merchandise, preferring to see it used rather than burned. Boston Police Chief Edward H. Savage

A self-propelled steam fire engine operates at the Great Boston Fire of 1872. Note the ruins in the background. The engine's hard suction hoses have been lowered into a secondary water source, an underground cistern.

and his men had great difficulty distinguishing between looters and property owners. Two firefighters were crushed to death when they attempted to rescue two men inside a burning store on Washington Street. The entire building collapsed.

Frequently, Damrell was called away from the fire lines to meet with city officials, including Mayor William Gaston, who demanded to know why the city was burning. Other officials were also insisting that Damrell begin to use dynamite to blow up buildings to create a firebreak. Damrell opposed the practice, but was forced to give in. After 2:00 a.m., explosions were heard around the city as officials—some with little skill—began blowing up buildings.

The fire raged through the night and into the next morning. It had consumed buildings that housed the *Boston Transcript* and the *Pilot*, Boston's Catholic paper, and destroyed the historic Trinity Church at the corner of Summer and Hawley Streets. Hundreds of thousands of dollars in leather goods, whole cloth material, paintings and rare books were lost as nearly one thousand firms were leveled. The fire also reached to the edge of Boston Harbor, which is now Atlantic Avenue, burning wharves and boats. At State Street, firefighters underwent a massive attempt to hold the fire from spreading. More than thirty steamers, many from out of state, grouped on State Street and poured water on buildings. Not until 1:00 p.m. was the fire deemed under control. However, some hours later, a gas explosion tore open a building on Summer and Washington

Aerial ladder work for rescuing victims from upper floors is a critical part of firefighting. This photo shows the steep angle of a wooden aerial ladder being used to rescue civilians on South Street, circa 1930.

Street and the blaze started again. Not until Monday morning was the Great Boston Fire finished.

More than sixty acres and 776 buildings were devastated. The damage in today's dollars would be close to $1 billion. Nine firefighters were killed fighting the "fire demon," as the inferno was dubbed. They were William Farry, foreman of Boston Hook and Ladder Company 4; Daniel Cochrane, assistant foreman, Boston Hook and Ladder Company 4; Henry Rogers, volunteer, Boston Engine Company 6; Lewis P. Abbott, ex-fireman and volunteer, Charlestown Fire Department; William S. Frazier, volunteer, Cambridge Fire Department; Frank D. Olmstead, volunteer, Cambridge Fire Department; John Connelly, member, West Roxbury Hook and Ladder Company 1; Walter S. Twombly, member, Malden Hose Company 2; and Thomas Maloney, member, Worcester Fire Department. Two more firemen later died from their injuries: Martin Turnbull, member, Charlestown Hose Company 3; and Albert C. Abbott, former member and volunteer, Charlestown Fire Department. The exact civilian death toll was never determined, but it was probably sixteen to thirty people.

A special fire commission was set up after the fire to investigate the cause and spread of the fire. Damrell came in for criticism for his handling of the fire, but the real cause was the unsafe conditions that he had warned the city about. John Damrell went on to found the National Association of Fire Engineers, which is now the International Association of Fire Chiefs. In 1877, he was appointed Boston's first building commissioner. He would hold that position for twenty-five years.

Today there is no evidence of the fire and its devastation, but a sign at 83–85 Summer Street marks the start of the fire. Boston rebuilt quickly, and even longtime residents are not aware of the terrible destruction in November of 1872. However, Boston firefighters have never forgotten the lessons of the fire of 1872. (For more on the 1872 fire, see entries on Old City Hall, Box 52 and Old South Meeting House.)
[S.S.]

6. Fire Alarm Box 52

At the intersection of Summer, Lincoln and Bedford Streets,
Downtown/Financial District

In the evening of November 9, 1872, a policeman cranked the first alarm of the Great Boston Fire from Box 52. In the 1870s, fire alarm boxes were actually locked to prevent the transmission of false alarms. People noticing a fire would have to get a policeman or a local merchant who had been provided with a key. Box 52 has retained a certain significance in the history of the Boston Fire Department; while it has been renumbered as Box 1431, the box is still marked as 52 on its outside. In 1873, the Box 52 Society, a group of Boston "sparks" or fire buffs, was formed to keep alive issues raised by the Great Fire. The association eventually disbanded, but was reorganized on the fortieth anniversary of the huge fire. Today members continue to host events on fire-related issues and to hold a banquet annually on the anniversary of the Great Fire. If you walk

Famous Box 52 was struck for this destructive general alarm fire on Thanksgiving Day, November 28, 1889. Chief of Department Lewis P. Webber views the ruins of Engine Company 26's steamer.

through the neighborhood in the vicinity of Box 52, you will see many brick and granite buildings with mansard roofs; they were built immediately after the fire and are similar to those that were standing in 1872.
[S.S.]

7. Thanksgiving Day Fire

69–87 Bedford Street, at the corner of Kingston Street,
Downtown/Financial District

Box 52 is known for its role in the Great Boston Fire of 1872. But the box also played a role in an equally destructive fire that came to be known as the Thanksgiving Day Fire. The weather on the morning of Thanksgiving Day, November 28, 1889, was terrible; there was a drenching, driving rain, with a strong southeasterly wind blowing at twenty-six miles an hour. Sometime in the morning, a fire started in a six-story granite building at 69–87 Bedford Street at the corner of Kingston Street. A police officer who spotted the fire sounded the alarm from Box 52 at 8:13 a.m.

The building at 69–87 Bedford Street was known as the Brown-Durrell Building, and it had a frontage of 107 feet by 162 feet. The fire was believed to have started in the offices of the Boston Electric Time Company, which regulated a number of electric clocks throughout the city. The police officer saw dense black smoke coming from the fifth and sixth floors of the Columbia Street side of the building after hearing a cry of "Fire, Fire!"

Engine Company 7 arrived promptly from quarters on East Street, just a short distance away, and started a line in the rear of the building. They forced the door, which inadvertently gave the fire a draft of air via the elevator shaft. The fire roared up the shaft with a savage intensity, and windows were blown out on the upper floors. Members of Engine 7 and Ladder Company 8 managed to unfasten the iron gate of the fire escape tower and carry a line of hose up to the burning fifth and sixth floors. They played a line of hose for about five minutes into the sixth floor, until they were driven back to the fifth floor. They were then notified that the fire was burning below them, so they were ordered down. Despite the efforts of other arriving companies, the fire began to spread across the narrow alleys to other buildings. Most of the buildings destroyed or damaged by this fire had been built after the Great Fire of 1872. Even with the heavy rain, the fire had plenty to feed on, and the rain did not seem to help the firefighters.

Fearing a repeat of 1872, Chief of Department Lewis P. Webber requested assistance from other departments. About forty minutes after the fire started, the building of origin collapsed. Pieces of the granite exterior blew off in flying chunks, severely hindering firefighting efforts.

Firefighters worked hard to prevent the fire from extending east of Columbia Street. After two hours of very hard work, the fire was stopped from spreading farther south, but it was not under control. Webber ordered several out of town steam engines to operate along Chauncey Street and Exeter Place as the fire moved along Bedford Street.

Assistant Chief John Regan relayed this story to reporters. During the early morning, he was on Kingston Street with Captain Knights of Engine Company 10 and Captain Frost of Engine Company 33 holding the spread of the fire at that location, with at least fourteen other firefighters. Without any warning, part of the wall of the building they were standing in front of came down. The men were knocked to the ground, but all averted death by some miracle. They picked themselves up and went right back to work. Newspapers reported that sixty-two steam engines responded and they used 107 streams on the fire.

The flames raged unchecked until they reached the Allen Building at the corner of Bedford and Chauncey Streets. Here and in the Nevins Building at Rowe Place and Chauncey Street, between 12:00 and 1:00 p.m., the firefighters were able to stop the fire.

On orders of Chief Webber, a detail of fifty men was held at the location to search for Laddermen Daniel J. Buckley and Frank P. Loker of Ladder Company 3 and Hosemen John J. Brooks and Michael Murnan of Hose Company 7, who had been caught in a collapsing building. All four died in the ruins.

After the fire, the fire department was strengthened greatly by the addition of several new engine companies, mostly in the downtown district. New buildings were built on

On the bitter cold night of January 28, 1966, a gas explosion resulted in a major fire in a downtown hotel in the "Combat Zone," the adult entertainment section of Boston. In a spectacular rescue, firefighters saved a woman from the basement via the blown out section of sidewalk.

the site of the fire, and the area became the nucleus of a more prosperous mercantile district.

[B.N.]

8. Paramount Hotel Fire

17–19 Boylston Street, Downtown/Financial District

It was Friday evening, January 28, 1966. The temperature was in the low teens with the wind blowing over forty miles per hour. In some firehouses dinner was over, and in others firefighters were waiting to eat. Friday nights always brought different attitudes in the workplace; some members were planning events for the weekend while others were just happy to be working. In 1966, you worked the two night tours together; some members would be off and others would be back on Saturday night.

About 6:38 p.m., a third-floor resident and part-time handyman for the Paramount Hotel, Herb McBride, detected an odor of natural gas in the stairway going down to the first floor. He brought this to the attention of the desk clerk, Ronald Coyne, and

Joseph Elliot, the elevator operator. The desk clerk advised him to notify the manager of Leonardi's bar, which was adjacent to the hotel lobby. McBride headed to the bar while Coyne walked toward the rear stairs and reached the first landing, where the odor was very intense. Alarmed, he started to retrace his steps to notify the gas company. Before he got very far, an explosion occurred. Coyne found himself in the cellar with Joseph Elliot.

The Paramount Hotel was located in a notorious area of Boston dubbed "the Combat Zone," which was then a seedy section of the Theatre District. On a typical Friday night, the area was packed. The eleven-story Paramount was part of a complex of buildings that were all about fifty years old, including the eight-story Plymouth Hotel, and other structures containing the Gilded Cage and Chartells Coffee Shop. At 6:38 p.m., Boston Police Officers George Ruck and William Crosby heard a loud explosion as they were patrolling the area. Police dispatchers alerted Fire Alarm Headquarters. Senior Fire Alarm Operator John McFadden received the call and Box 1471 was struck at 6:40 p.m. Engine Company 7, Ladder Company 17, Engine Company 26, Engine Company 10 and Ladder Company 8 with the rescue company, Deputy Chief John O'Mara of Division 1, District Chief Galvin of District 4 and Acting District Chief Feeney in District 3 all responded on the first alarm. O'Mara allegedly remarked to his aide John Donovan as they were en route, "I'm glad I have these heavy socks on in case this turns out to be something." It turned out that the additional footwear was really necessary.

When Engine Company 7 arrived at Box 1471, all the men could see was smoke, haze and dust. Engine 7 Captain Charles Griffin called Fire Alarm over the radio with a sense of urgency in his voice. "We're at the box, we don't know what we have." Engine 7 let Ladder 17 cover the front of the building. They could glimpse flames inside the hotel, so Captain Griffin ordered a big line started.

The street was now filled with screaming people; frantic hotel guests could be seen at windows on the upper floors, calling for help. One person was hanging over the balcony unconscious. The fire at the rear of the building intensified and started to spread to the upper floors and toward the front. Gas could be smelled in the street as the companies arrived. District 4 Fire Chief Galvin requested ambulances to come up Washington Street from Stuart Street. Deputy Chief O'Mara ordered five alarms struck.

The explosion had blown out part of the first-floor walls and the sidewalk was now in the basement. Ladder Companies 15 and 18 roared down Boylston Street from Tremont and were able to throw their "sticks," or aerial ladders, onto the front of the building.

Engine Company 26 ran a big line to the front of the building and started to hit the fire in the front while Engine 7 was doing the same. Ladder Company 17 threw a thirty-five-foot ladder to the extreme right side of the balcony level next to the Paramount Hotel sign. Ten occupants of the hotel safely escaped using this ladder. The chauffeur was getting the big Seagrave one-hundred-foot aerial up, as there were people waiting in the windows. As these companies were arriving, the police were removing a victim.

As the ladder companies arrived, they were ordered to get ground ladders, as the trucks could not get close enough. Companies were able to get into the small alley that ran on the "exposure 2" side, known as Bumstead Court. This gave access to this side of the building, where ground ladders were used. The cold temperatures turned the water from the hoses into ice almost at once. Extra manpower was needed with the ground ladders due to the severe weather.

One of the most spectacular rescues ever witnessed happened at this fire. A woman was seen in the basement area as the sidewalk was blown away. Heavy fire was showing in this area. Firefighter William "Bill" Shea of the Rescue Company jumped into the basement with total disregard for his safety. The woman was unconscious and pinned by a beam; only her head was out of the water pooling from the fire hoses. She was not only in danger of being burned, but also of drowning. Firefighter Shea managed to get the woman up, and with the help of other firefighters a ladder was dropped into the basement. Shea managed to pull the woman to safety. Firefighter Shea was burned about the ears and hands and was transported to a hospital; he would remain off duty for weeks. He was awarded the John Fitzgerald Medal for Most Meritorious Act of 1966.

The fire left eleven dead and dozens injured. "The magnificent courage of the firefighters, police and other city workers earned our gratitude," said Boston Mayor John F. Collins after he visited the scene.

Today, the notorious "Combat Zone" has almost vanished.

[B.N.]

9. Chinatown Fire

84 Essex Street, Chinatown
Thursday, October 12, 1989, was a typical beautiful fall day in Boston. It was the twenty-sixth anniversary of the Morgan Memorial Fire, and it was also Fire Prevention Week. A CBS production crew for the television magazine program *48 Hours* was in Boston filming part of a special on fire across the United States. Their expectations for a major incident were not high, as fire activity was down in most large cities.

At 11:52 a.m., Fire Alarm received the first of several calls reporting a fire in a building opposite 84 Essex Street in Boston's congested Chinatown section. Engine Company 10, the Tower Company and District 3 Chief Ronald T. Marston were "stilled," and Box 1432 was transmitted. Filling out the first alarm assignment were Engines 7 and 5 (from Engine 3's quarters), Ladders 17 and 15 (in place of Ladder 18), Rescue 1, the Squrt Unit and Division 1 Deputy Chief Paul A. Christian.

One minute later, Rescue 1, which had been on the Expressway returning from headquarters, reported they were on Essex Street with fire showing from the second floor. The fire originated in the six-story Edinboro Building, better known as Ming's Market. The area, once known as Boston's shoe and leather district, was densely packed with six-, seven- and eight-story buildings. Many dated back to the nineteenth century. By the late '60s, many had been converted from shoe and leather trades to Oriental markets and restaurants, light manufacturing and office space. Some of the buildings were equipped with sprinkler systems or fire alarm systems. Vehicular congestion was

Deputy Chief Paul Christian evaluates options during a spectacular fire in Chinatown on October 12, 1989. Nine alarms were struck on Box 1432 at Essex and Edinboro Streets. A CBS-TV production crew for *48 Hours* videotaped this fire.

always a problem in the area, with double-parked vehicles frequently reducing Essex Street to one lane.

The first arriving companies ran hand lines into the building in an attempt to cut off the upward extension of the heavy fire. Rescue 1 began evacuating occupants from the building.

The first due ladder company, Ladder 17, finding the sprinklers inoperative, went to the basement to check on the system and found the sprinkler piping dismantled. On the arrival of Chief Marston, fire was rapidly taking possession of the entire second floor and was spreading back along the Edinboro and Ping On Street sides of the building. Chief Marston ordered the interior attack abandoned and heavy-stream appliances put into operation. Deputy Chief Christian, arriving moments later, sized up the situation and ordered the second alarm struck.

For the next four hours, the Boston Fire Department would be engaged in an epic struggle, hampered by water supply problems and illegal detonating fireworks, to contain a raging inferno in several mercantile buildings.

All first alarm engine companies reported difficulty in water supply from the usually reliable high-pressure fire system. Chief Christian ordered the high-pressure system pressure increased to 125 pounds per square inch. Still there was no improvement in the water supply; the problem was not at the pumping station. Resourceful pump operators began connecting century-old Lowry hydrants in the street.

At last a water supply was secured. But this was not enough. Flames had spread into the adjacent eight-story building to the right of the original fire building at 81–83 Essex Street, and in the rear the fire was extending down Edinboro Street.

Heavy-stream appliances were the order of the day, and responding engine companies were ordered to connect to post hydrants and lay in with feeder lines to the companies closer to the scene. Alternative water supply had to be established and plenty of water was needed—fast.

Multiple alarms were struck in rapid succession. At 12:07 p.m., the deputy chief struck the eighth and ninth alarms.

Companies responding on the multiple alarms were ordered into Beach Street to attack the fire from the rear and into Edinboro Street to stop extension in that direction. Adequate water was secured on Beach Street, and the rearward extension of the fire was halted after a determined effort by Boston firefighters.

Fire Commissioner Leo D. Stapleton arrived and assumed command. A collapse zone was established and apparatus were pulled back in preparation, should the buildings collapse. Ladder pipes and aerial tower streams were put in readiness and hose lines were dragged in from distant water sources.

Interior operations were attempted in 81–83 Essex Street from the safety of the fireproof stairwells, but the intense exposure proved too much and firefighters had to retreat to an exterior attack. Ladder pipes and deck guns were put into operation. This proved fortuitous, for moments after the interior attack was abandoned several of the top floors flashed over, the granite façade of the building began "spalling"—showering the firefighters with chips of granite—and a continuous stream of loud explosions began emanating from the upper floors. Suddenly, large bundles of unexploded fireworks began dropping to the street. For the next hour, firefighters not only had to contend with battling flames, but exploding fireworks as well.

With adequate water supply established and an aggressive exterior attack facing intense heat and thick smoke, the operations were successful and the fire was contained to 81–83, 85–91 Essex Street and 1–9 Edinboro Street. After the heavy fire was knocked down and the explosions ceased, interior operations were resumed; a major blaze had been subdued after Herculean efforts by Boston firefighters.

Investigation revealed that the fire was the result of contractors using a torch (illegally) inside of an occupied building to (illegally) remove the sprinkler system piping. After the fire broke out, they fled without sounding an alarm. The high-pressure fire system was apparently damaged by an unscrupulous contractor who shut the gate valves on the Essex Street branch and neglected to report it to authorities. Investigators in 81–83 Essex Street also discovered a large cache of illegal fireworks.

Ladder work has always been a hallmark of Boston firefighters. In this view, firefighters advance hose lines over ground ladders and aerial ladders to the upper floors of the fire building.

This popular dining and dancing establishment, the Pickwick Club, collapsed in the early morning of July 5, 1925, killing forty-four persons. No fire resulted, but the fire department was called in to conduct rescue and recovery operations.

No fatalities or serious injuries were reported, thanks to the strenuous efforts of firefighters to contain this fast-moving inferno.

The CBS *48 Hours* crew had not only seen Boston firefighters at their finest, but also produced the best segment of the nationwide presentation "Fire in America."
[P.C.]

10. Pickwick Club Collapse

6–12 Beach Street, Downtown/Financial District

One of the city's worst disasters occurred on July 5, 1925, in the Pickwick Club, which was located in a brick building on the north side of Beach Street in Boston's Theatre District. Just a few doors east of Washington Street, the Pickwick Club was a private social club that was open to members only; it occupied the second floor of the five-story building. The first floor was occupied by the Greenwich Village restaurant, and the other floors were vacant.

At the time of the collapse, the Shoppers Garage was under construction on the east side abutting the Pickwick Club building. On Saturday evening, July 4, 1925, about 125 people were in the building, celebrating the Independence Day weekend.

Shortly before 3:00 a.m., with the crowd pounding the dance floor, the entire building suddenly collapsed into the cellar, trapping many persons in the debris. Fire Alarm Box 1471 at Washington and Essex Streets was pulled and transmitted at 3:01 a.m. on Sunday, bringing the first alarm assignment of four engine companies, two ladder companies, the water tower and a rescue company to the scene. Firefighters worked all night with police, ambulance personnel and volunteers, pulling survivors and victims from the wreckage. Many were trapped for up to eight hours, and the fire department was involved only in a rescue operation, as there was no ensuing fire. Police sent ambulances to the scene and more were sent from the Boston City Hospital. The Boston Edison Company also responded and set up large floodlights to aid in the rescue operation. The fire department remained on duty for sixty hours, and finally left the scene at 10:45 a.m. on Tuesday, July 7. The final death toll was forty-four.

A fire in the same building had been recorded several months earlier on Monday evening, April 23, 1925. A close examination was made after the building collapse in July, but no connection between the two events was ever proven.

To investigate the collapse, a grand jury was convened; they concluded that the easterly wall of the Pickwick Club building, which had been exposed because of the new building construction next door, had not been properly shored. When the wall collapsed, it brought all the floors with it to the basement. The Shoppers Garage is still standing today, and a new high-rise tower occupies the site of the collapsed building.
[J.T.]

The firefighters and motor apparatus of Engine 7 pose in front of their new firehouse at 7 East Street, downtown, circa 1923. This station is one of only a few Boston firehouses to have the Boston Fire Department name inscribed above the apparatus doors. It replaced the firehouse built in 1870.

The firefighters and equipment of T.C. Amory Engine 7 pose outside the company's new firehouse at 7 East Street, downtown, after it opened in 1870. Fire companies at that time were identified by both a name and a number.

11. Former Engine Company 7 Quarters

7 East Street, Downtown/Financial District

This old single-door, three-story firehouse was the quarters of Engine Company 7 from 1923 to 1953. The building at 7 East Street replaced an older firehouse at the same location that was built in 1870. Steam Fire Engine 7 was organized on January 1, 1859, in a firehouse on Purchase Street near Congress that had originally been the quarters of a hand-drawn company. The company was equipped with a steam fire engine built by Bean and Scott of Lawrence, Massachusetts, and was originally operated under contract by the builder. The original firehouse was inadequate for a horse-drawn steam engine, and the city built the new firehouse at the corner of East Street and East Street Place and moved Engine 7 to this location in January of 1870.

During the Great Boston Fire of November 9–10, 1872, Engine 7 was the closest engine to Box 52, which was struck for the big fire. There was a long delay in sounding the alarm, and the company responded from East Street at about 7:15 p.m., before the bell alarm struck in the firehouse. They were alerted by loud noise and a commotion coming from a crowd in the street that was watching the fire two blocks away on Summer Street. Because the horses were disabled by the epizootic epidemic, the engine was drawn by a crowd of one hundred men and boys to Kingston Street. Engine 7 was the first engine on the scene and they got to work immediately, playing water from their line on the rapidly spreading fire in the alley at the rear of the fire building on Kingston Street.

After fifty-two years of service as one of the city's busiest fire companies serving in the hazardous leather district of South and Lincoln Streets, the fire horses were retired. To meet the needs of the motor era, the old house was torn down and a modern new firehouse was built at the same location. It opened on June 27, 1923, and still stands today on East Street.

On November 12, 1953, the East Street firehouse was closed and Engine 7 moved to the quarters of Engine 25 and Ladder 8, a half-mile away at 123 Oliver Street. The company moved again on July 13, 1954, to the quarters of Engine 26 and Ladder 17 at 194 Broadway. When that house was razed for redevelopment in 1971, Engine 7 made its latest move to 200 Columbus Avenue in the South End on May 10, 1971. The East Street firehouse was occupied for many years by a commercial window company, and was recently sold and renovated into a restaurant.

[J.T.]

12. Sleeper Street Fire

23–27 Sleeper Street, near Congress Street, South Boston

During 1947, the Boston Fire Department made great progress in getting up to full strength following personnel shortages created by World War II. Early in the year, the nature of the job changed radically with the replacement of the eighty-four-hour week, two-platoon system with the forty-eight-hour, seven-group system. New men were hired, many of whom were veterans of World War II. At the same time, the apparatus fleet began to be modernized. Fire hazards, however, did not abate.

The Boston Wharf Company at that time owned blocks and blocks of dry goods warehouses in the northwest corner of South Boston, known as the "wool district." One of its tenants was the Armour Leather Company at 23–27 Sleeper Street.

On January 15, 1948, the department responded to twenty-three box alarms, three of which required working fire responses, while four more resulted in multiple alarms. Snow covered the ground and the average temperature that day was fifteen degrees. Shortly after 12:40 p.m., the Fire Alarm Office received signal 222-416-8 from the Boston Automatic Alarm Company, a private fire alarm agency, and transmitted that signal to Boston's firehouses at 7:48 p.m., followed by Box 7115 one minute later.

Four more alarms for 23–27 Sleeper Street followed. Companies responding from the nearby engine and ladder houses attempted an interior attack on the Boston Wharf Company complex, but soon were forced to relocate. Huge clouds of smoke banked down onto Sleeper Street and spread throughout the warehouse district near the Fort Point Channel.

Conditions deteriorated throughout the afternoon so that recall signals were transmitted at 4:30 p.m. This meant that not only were the working groups kept on duty at the fire, but other groups that were not scheduled to work until 6:00 p.m. were also called.

This is a view of the Sleeper Street Fire on January 15, 1948, looking from Northern Avenue toward Congress Street. Sleeper Street is an extremely narrow street located in what was known as the Boston Wharf Company Warehouse District.

An 1896 Amoskeag steam fire engine, attached to a 1917 Christie motor tractor, is returned to service to assist in thawing ice from apparatus and equipment the day after the Sleeper Street Fire. Ladder 18's aerial snapped under a heavy accumulation of ice and is shown at the right side of the photo.

All the horse-drawn fire apparatus of the double Engine Company 38/39 are on display in front of their quarters at 344 Congress Street, circa 1894. This station was active for eighty-six years, from 1891 to 1977, and is now the home of the Boston Fire Museum.

The subfreezing temperatures created dangerous conditions as spray from the hoses froze on the streets, building façades, apparatus, tools, helmets and coats. This fire was a "long stand," with the all out at 10:52 a.m. on the following day.

In an unusual move, an Amoskeag steam fire engine drawn by a 1917 Christie front-drive gasoline tractor was ordered to the fire and used to thaw out equipment and frozen hose. The wood aerial ladder of Ladder 18 snapped in two due to the excessive weight of accumulated ice.

Today the area is in multiple ownership, and is being preserved by new owners who highly value these handsome old warehouses. After the fire, the remains of Armour Leather were torn down and the lot is still vacant; but nearby properties have been remodeled into lofts, condominiums, offices and other contemporary uses. The whole district—which at one time was populated only by warehouse workers and truckers and was all but deserted after 5:00 p.m.—now is filled with office workers and residents by day and restaurant- and club-goers in the evening.
[T.G.]

The Congress Street Fire destroyed several downtown buildings on Congress Street and Federal Street on March 9, 1950. Engine 25 and Ladder 8 found the structures fully involved upon arrival at 3:00 a.m. The Bank of America building now occupies the site.

13. Congress Street Fire Station

The Boston Fire Museum, 344 Congress Street, South Boston

The Congress Street Fire Station was built in 1891 and was the site of the only fire company in Boston Fire Department history to be organized as a double-engine company. Engine Company 38–39 was organized on May 18, 1891, with Captain John F. Ryan in command. It was deemed necessary to build a fire station in the area because of the abundance of wool warehouses in the neighborhood and the extensive rail yards and the Fort Point Channel surrounding the district. The rail yards and the channel isolated the area from the rest of the city. The bridges over the channel, which was navigable until the 1950s, frequently opened, and rail cars frequently blocked streets that were crossed by the railroad tracks. In case of a fire in an area that was blocked off, the next nearest fire station was at Broadway and Dorchester Avenue, nearly a mile away. The Boston Wharf Company, principal land and building owner in the area, arranged to sell the land to the city for $5,197.80 for the express purpose of building a fire station. The building was designed by city architect Harrison H. Atwood using a combination of Romanesque and panel brick styles. The contractor, Connery and Wentworth Co., was paid $36,292.27 to construct the building. The building is trapezoidal in design, measuring thirty feet wide in the front and forty-six feet wide in the rear. In the interior, the building is a "hanging building," meaning the second floor hangs from the roof by a series of iron columns connected to the truss roof, thus freeing the apparatus floor of any pillars. A line of horse stalls across the rear of the apparatus floor housed the horses needed to pull all the apparatus in the station. In 1897, Engine 38's horse-drawn steam pumper was replaced by a self-propelled steamer, one of only two in the city, and continued in operation until 1925. The other horse-drawn apparatus, Engine 38's hose wagon and Engine 39's steam pumper and hose wagon, were replaced by motorized apparatus in 1917. Engine 38 was deactivated in 1947, and in 1953 Ladder Company 18, which was quartered in a fire station at 9 Pittsburgh Street (about seventy-five yards away), moved in.

The fire companies in this station responded to many serious fires through the years. Principal among those is the Merrimac Street Fire of 1898, in which five members of Engine Company 38–39 were killed: Captain James H. Victory, Lieutenant George J. Gottwald, Hoseman Patrick H. Disken, Hoseman John H. Mulhern and Hoseman William Welch.

The fire station was closed on April 22, 1977. In 1979, the station was dedicated to Boston Pops conductor Arthur Fiedler as part of the Transportation Museum. In addition to his musical skills, Fiedler was also a devoted "spark" who often rode with fire engines to working fires.

In 1983, the Boston Sparks Association acquired the building and organized the Boston Fire Museum. In 1987, the Congress Street Fire Station was approved for inclusion in the National Register of Historic Places. The Boston Fire Museum contains antique Boston fire apparatus, collections of artifacts and memorabilia and many pictures and photographs of Boston fires and firefighters. Principal among the exhibits is a display on the Cocoanut Grove Fire, containing chairs, plates and other small items

The popular nightclub Blinstrub's was the victim of a major fire on February 7, 1968. The club was located on the corner of West Broadway and D Streets in South Boston and showcased many popular stars and singers, including Tony Bennett and Dean Martin.

salvaged from the Grove after the fire. Another important display is the small hand tub, "Boston 1," a 1792 Thayer that served in Boston for many years. The museum is usually open on weekends in the summer months.
[M.G.]

14. Blinstrub's Nightclub Fire

D Street and West Broadway, South Boston
Shortly before 9:00 a.m. on the cold crisp morning of February 7, 1968, fire broke out in the basement of Boston's last big-time nightclub, Blinstrub's Village (known to locals as Blinnies). Located at D Street and West Broadway in South Boston, the nightclub was New England's largest, and it featured many of the day's biggest stars: Wayne Newton, Patti Page, Dean Martin, Tony Bennett, Jimmy Durante, Liberace and the McGuire Sisters, to mention a few. The two-story building, with a Bavarian castle façade, had been expanded over the years and occupied a prominent portion of the block.

Smoke was discovered in the basement and Box 7222, at D Street and West Broadway, was pulled at 8:44 a.m. Little smoke was showing when the first companies arrived, but a stubborn blaze was underway in the basement. The crew of Engine 39 ran a big line into the basement through a door on D Street, only to be driven out when the fire advanced up through the structure.

A working fire was reported by District Chief James Murphy at 8:53 a.m., and Deputy Chief George Thompson ordered a second alarm on his arrival at 8:59 a.m. Fire raced through the old wooden structure, forcing firefighters into the street after a narrow escape. Thick, black smoke surged through every opening in the building and filled the South Boston sky with an ominous plume. A third alarm was ordered at 9:07 a.m., followed by a fourth at 9:13 a.m.

Flames engulfed the entire club and threatened the congested neighborhood. Boston Police Station 6, located directly behind the club across narrow Athens Street, was also threatened. Concerned police officials removed ammunition and guns from the building as a precaution.

News of the fire spread rapidly and the street was soon filled with onlookers, including owner Stanley Blinstrub and Boston's new mayor, Kevin H. White.

Acting Chief of Department Francis X. Finnegan took command at 9:16 a.m. and ordered a fifth alarm sounded. Hoses were set up to cover exposures and deck guns were directed through the many small windows into the club. Hose lines were directed from nearby rooftops. Throughout the morning and into the afternoon, firefighters fought ice, smoke and cold to contain the inferno.

By late afternoon, Blinnies had staged its last performance; it lay in ice-encrusted ruins. At 5:00 p.m., a large dump truck with shotgun-armed Boston Police arrived on D Street and stood by as Stanley Blinstrub directed the removal of the safe from the ruins.

Plans were made for a new Blinstrub's in Dorchester off the Southeast Expressway near Freeport Street, but time passed and the project never developed. An era had ended in Boston with the destruction of its last major nightclub. Names like the Latin Quarter, the Mayfair, Cocoanut Grove and Blinstrub's now belong to history.

Today, the intersection of D Street and West Broadway is quite different. A gas station occupies the Blinstrub's site, and Police Station 6 was closed in the early '80s. A fast-food restaurant now occupies the corner lot to the north of Blinstrub's, and in 1977 the D Street firehouse, opposite Station 6 on D Street, was opened.

Blinnies lives on in the memories of those Bostonians who lived through that golden era of live big-name entertainment.

[P.C.]

Other Boston Areas

1. Firefighter Memorial

Firemen's Lot, Forest Hills Cemetery,
95 Forest Hills Avenue, Jamaica Plain

The 250 acres of the Forest Hills Cemetery, established in 1848, are a bit off the beaten path for Boston visitors, but the cemetery is well worth a visit for its lovely garden design, its Victorian and modern sculptures, its art installations and one of the region's most moving memorials to firefighters.

The Charitable Association of the Boston Fire Department bought the Boston Firemen's Lot in the Forest Hills Cemetery in December of 1856. The association had been formed in about 1828 to help the families of firefighters who were killed or injured on the job and to ensure that indigent firefighters received a decent burial. According to the deed, the lot contains 13,594 square feet divided into 252 gravesites and holds the remains of 135 members, 13 of whom were line-of-duty deaths.

The first two burials at Forest Hills Cemetery were of Francis F. Cutting, age twenty-five, and John W. Tuttle, age thirty-six, both assigned to Tremont Company 12. They lost their lives on May 2, 1858, at 133–139 Federal Street when they were crushed by falling walls.

A twenty-six-foot-high monument was later commissioned for the site as a permanent marker for the 117 firefighters who were then buried in the plot. The monument consists of a nine-foot fireman cast in bronze set on a fourteen-foot-high Quincy granite base. With his head turned to the left, the figure stands proudly in turnout gear, hose line belt and helmet. The figure, reportedly modeled on a firefighter named Cosgrove, was sculpted by John Albert Wilson (1878–1954); the base was prepared by White & Sons of Quincy. Each side of the pedestal has a bas-relief bronze plaque that shows a different scene of historic firefighting: a horse-drawn steamer engine racing to a blaze; a hand tub; a hook and ladder truck; and a protective wagon.

About five thousand people, including officials, firefighters and civilians, attended the memorial's dedication in June of 1909. Among them was former Boston Mayor John

The Boston Firemen's Lot at Forest Hills Cemetery is dominated by an imposing firefighter statue. Dedicated in June of 1909, the twenty-six-foot-high monument stands vigil over the final resting place of members of the Boston Fire Department.

Francis "Honey Fitz" Fitzgerald, the grandfather of future President John F. Kennedy. Nathaniel Taylor of the *Boston Globe* remarked in a speech at the dedication,

> *The many noble firemen who served the City of Boston efficiently and honorably did the best they could for the public good. The saving of life was their province. That they did all that was possible for human beings to do in their line of duty is the sincere belief of all our citizens. To the departed we say rest in peace. To their living comrades we say your life work is thoroughly appreciated by the City of Boston.*

The monument was then unveiled by Margaret and Josephine McLean of Ashmont, the young daughters of Captain Walter McLean of Engine Company 46, who had served as chairman of the monument committee.

One of the plaques was stolen off the monument some years ago, but was recovered after cemetery staff publicized a photograph of the plaque. A potential buyer, who saw the plaque in a New Hampshire flea market, found the photo during his research and alerted cemetery staff.

Opposite the memorial is the grave of John Stanhope Damrell, who was chief of the fire department during the Great Boston Fire of 1872. Another famous firefighter buried in Forest Hills is Fire Chief William T. Cheswell, who died at age sixty-three of a heart attack on February 15, 1906, while fighting a fire on Commercial Street in Boston's North End. The last firefighter to be buried here who was killed during a fire was Firefighter Edward R. Connolly, who was killed by a building collapse on March 21, 1986, at a fire at Box 7413. Connolly was past treasurer of the Charitable Association. Firefighter David Middleton of Engine Company 51 was buried here in June of 2007. Middleton died of a heart attack after working two fires the previous night, so his death is considered a "line-of-duty" death.

The second Sunday of every June, the Boston Fire Department holds an annual memorial service in the Forest Hills Cemetery to honor all those who have given their lives to the protection of others. The ceremony, which the public may attend, begins with a morning Mass at the cemetery's chapel, then a slow march with bagpipes to the Firemen's Lot, where flowers are laid on the graves of fallen firefighters. On hand are ladder trucks with aerial ladders raised to hold up a huge American flag. Visitors may make the march along with firefighters and their families and pay their respects.

The cemetery grounds are open every day from 7:30 a.m. to dusk. It can be reached via the Orange Line T to Forest Hills Station or via the VFW Parkway by car. The cemetery's website has detailed driving directions.
[P.C., S.S.]

2. Luongo Restaurant Fire

12–16 Maverick Square at the corner of Henry Street, East Boston

The Luongo Restaurant Fire, also known as the Maverick Square Fire, is hardly known by the public but is well known to Boston firefighters as one of the city's worst fires. Why

Firefighters remove the body of a brother firefighter from the rubble after the collapse of the Lyceum in the Maverick Square fire of November 15, 1942. Six firefighters died at this fire, an event that was soon overshadowed by the Cocoanut Grove nightclub disaster. Ladder 8's truck in the foreground was severely damaged, but was later rebuilt.

was it so terrible? On November 15, 1942, six firefighters were killed, forty-three injured and many others were trapped under debris for up to eighteen hours. The story of the East Boston fire was pushed off the front pages by the terrible loss of life at the Cocoanut Grove only two weeks later on November 28, 1942.

The fire started in the rear of Luongo's Restaurant, which was located in the first floor of the one-hundred-year-old Maverick Lyceum, across the street from the Maverick Square MBTA station. A night worker discovered the fire, which was probably due to an electrical appliance. He ran into the street and enlisted the aid of two citizens, who tried in vain to fight the fire. The fire alarm office received a call reporting the fire at 2:26 a.m. At 2:27 a.m., the East Boston companies were notified and responded from their fire station a few blocks away. The second alarm was sounded at 3:04 a.m., when the fire seemed to be making headway. The third alarm was sounded at 3:24 a.m.

Then at 4:15 a.m., without warning, the wall on the Henry Street side of the building bulged and collapsed, trapping firefighters in the building and burying Ladder 8. Fourth and fifth alarms were sounded to help in the rescue effort and to fight the fire, which had started up again after the collapse. Within half an hour ambulances, doctors, additional firefighters and rescue workers (including the coast guard) were rushing to the scene.

The Luongo Restaurant Fire of November 15, 1942, is also known as the Maverick Square Fire. Lyceum Hall in Maverick Square, a prominent social landmark, was destroyed by the fire.

The scene was one of pandemonium. Firefighters who were lucky enough to be clear of the wreckage, although injured, made desperate efforts to save their comrades. Smashed in the collapse was the new American LaFrance 125-foot steel aerial ladder of Ladder Company 8, the largest in the nation at that time.

Meanwhile, the flames were crackling high in the air. The adjoining building on Henry Street caught fire and some fifty persons fled in their nightclothes. When daylight came, exhausted and injured firemen were lying on the street, waiting to be removed to hospitals. The quick arrival of doctors and ambulances saved many lives.

The firefighters who died were on the second floor working with hose lines. They were Hoseman John F. Foley, Engine Company 3; Hoseman Edward F. Macomber, Engine Company 12; Hoseman Peter F. McMorrow, Engine Company 50; Hoseman Francis J. Degan, Engine Company 3; Ladderman Daniel E. McGuire, Ladder Company 2; and Hoseman Malachi F. Reddington, Engine Company 33. At Boston City Hospital, there were rows of injured firefighters with broken limbs waiting for X-rays. They did not complain, as they were the lucky ones. The saddest part of the tragedy was that the fire was "under control" when the collapse occurred. On a rainy Tuesday, November 24, 1942, Boston firefighters, still in shock, gathered at Holy Cross Cathedral for a memorial Mass for their fallen comrades.

[B.N.]

3. Roxbury Conflagration

Walpole and Berlin Streets, Roxbury

Early May of 1894 was a hot and dry period with hazardous fire conditions in New England. Major forest fires were being fought in Maine and Rhode Island, and there had been no rain for a while. Tuesday, May 15, 1894, was actually a bit cooler, with a high temperature that afternoon of sixty-five degrees in downtown Boston and a recorded wind of twenty-four miles per hour blowing from the northwest.

It was 3:00 p.m., and Boston's champion baseball team, the National League Beaneaters, was playing ball at the South End Ballgrounds, located at the end of Walpole Street next to the railroad tracks. This was their twenty-fourth year, and the season had opened on April 19 with a game against Brooklyn before a crowd of eight thousand. The ballpark was one of the grandest in the league and featured the Grand Pavilion, which had a two-story grandstand built in 1888. It was a tall structure and very ornate, with turrets and spires at each end reaching to the sky, and all built of wood. The Beaneaters had won the National League pennant in 1891, 1892 and 1893. They were falling behind in the 1894 season, but continued to draw big crowds, especially this day, as they played the Baltimore Orioles. Just to the east of the ballpark was a crowded neighborhood of old one- and two-family wooden houses that had been built before Roxbury was annexed to the city in 1868. Columbus Avenue had not been extended south of Camden Street at that time; the nearest major street was Tremont Street one block to the east. Tremont Street was a wide thoroughfare and contained some large brick and wood apartment blocks, as well as the brick firehouse that was the quarters of Hose Company 7 and Ladder Company 12.

At the end of the third inning, a spectator sitting in the right field bleachers lit a cigarette and dropped the match through a crack and into a pile of rubbish underneath the seats. This started a small fire that took hold in some rubbish, and it quickly spread along the bleachers. The game was halted temporarily, and some of the players made an effort to extinguish the blaze. The police officer on duty at the park ran to the firehouse on Tremont Street and found that the companies were already starting to the fire. Box 215 at Tremont and Cabot Streets was transmitted at 4:22 p.m. The strong winds carried the fire into the grandstand, and soon the entire structure was in flames. The crowd quickly evacuated into center field and watched in amazement as the fire ignited houses on Berlin Street and then spread along Walpole Street toward Tremont Street. The fire engulfed houses on Walpole, Cunard, Burke and Coventry Streets, but then stopped at Tremont Street, which was a wide firebreak.

Fire companies responding to the first alarm from Box 215 encountered difficulty, as there were no hydrants on the streets west of Tremont leading to the ballpark. Long lines of hose had to be stretched, which consumed much time, and this delay was a factor that caused the fire to spread very quickly. Box 98 at Massachusetts and Columbus Avenues was struck at 4:25 p.m. and acted as the second alarm. The third alarm from Box 215 was sounded at 4:42 p.m., and the fourth alarm at 4:51 p.m.

The steamers of Engines 24 and 42 were heavily damaged and 110 buildings were destroyed in the Roxbury Conflagration on May 15, 1894. The fire started in the grandstand of the Walpole Street Ballgrounds. Note the Lowry hydrant feeding the two steamers.

Engines 24, 42 and 37 were all connected to a hydrant at Tremont and Cunard Streets and made a stand at that location to prevent the fire from jumping Tremont Street. The fire then unexpectedly built up headway at this location and jumped over the fireman and carried the fire toward Cabot Street. The engines had to be abandoned due to extreme heat, and all three suffered extensive damage, but were repaired and returned to service after the fire. At the height of the fire, the Fire Alarm Office received and transmitted additional alarms for roof fires on Ruggles Street, Williams Street, Shawmut Avenue, Dudley Street and Greenville Street, a half-mile to the east. The main fire was eventually contained at Cabot Street.

There were no serious injuries that day, but the final count showed that 110 buildings were completely destroyed and 106 structures were extensively damaged. The Pavilion and other structures owned by the Boston Baseball Association were a total loss, with damage of $70,000. The city of Boston lost a brick schoolhouse at Tremont and Walpole Streets, as well as the firehouse of Hose 7 and Ladder 12. The fire companies were temporarily relocated to Engine 13 quarters at Cabot and Culvert Streets. A new firehouse was built on the same site at 1046 Tremont Street, which opened in May of 1896. Twenty firemen lost personal items at the old firehouse, and these were compensated by the city. In addition, hundreds of poor families with no insurance lost their homes and personal belongings.

The baseball team moved temporarily to the Congress Street ball grounds in South Boston. The Walpole Street grounds were quickly rebuilt, but on a smaller scale with a single-level grandstand. The new park reopened on Friday, July 20, 1894, with a game against New York that drew a crowd of 5,206 spectators.

[J.T.]

4. Plant Shoe Factory Fire

Centre and Bickford Streets, Jamaica Plain
The massive complex that once housed the former Thomas G. Plant Shoe Factory was the site of a spectacular five-alarm fire in the late evening of February 1, 1976. The complex, built during the late 1890s and early 1900s, was first used as a shoe factory, then by several companies in shoe-related businesses and later by artists, crafts shops and small-scale manufacturers. The factory was a major source of employment for many German and Irish immigrants living nearby.

At 9:28 p.m., the Fire Alarm Office sent Engine 14 (which was returning from another alarm) from Dudley Street, Ladder 10 from Centre Street and the Chief of District 9 Leo McElaney to a reported fire at 301–309 Centre Street, Jamaica Plain, followed by an alarm struck for Box 2411 one minute later. No alarms were skipped, nor were the spacing of the second through fifth alarms indicative of extreme urgency; yet the fire grew to such magnitude that many special calls followed the fifth alarm as the complex became involved "stem-to-stern."

The severity of this fire cannot be appreciated unless you can visualize a series of two- to six-story buildings approximately 680 feet by 240 feet with fire showing from every window. The fire drew hundreds of spectators from the congested Jamaica Plain neighborhood and eventually from all over Greater Boston as forty-three engine and twelve ladder companies, including several mutual aid companies, responded.

Early in the progress of the fire, steady rain fell, with temperatures ranging in the unseasonably high forties and fifties. Spectators standing on the southerly side of Centre Street opposite the five buildings observed Fire Commissioner George H. Paul glancing alternately at the advancing flames and at the row of three-deckers behind him. Paul ordered several heavy-steam appliances, including two aerial towers, to Centre Street to keep the fire contained to the Plant building. In fact, the heat generated by the fire became so intense at one point that the falling rain evaporated before reaching the ground and caused it to steam off the wooden three-deckers.

The fire department struggled through the night with the fire as several sections of the huge, heavy-timbered mill building collapsed. During the morning and early afternoon of February 2, the temperature dropped steadily, and with strong winds developing, the effective temperature reached negative forty degrees. As a result, several pieces of fire apparatus were disabled, and many personnel were covered in a heavy layer of ice.

The movement of fire companies to the fire and ensuing coverage by out of town apparatus in the downtown also created a potential problem: only Boston engine

This view shows firefighters operating a deck gun to protect surrounding properties from the heat generated by arguably the largest single building fire in the city's history. This spectacular fire occurred on February 1, 1976, in the 700- by 250-foot Plant Shoe factory.

companies carried the specialized keys and wrenches needed to operate the high-pressure hydrants in the downtown area. Accordingly, Fire Alarm ordered Engine Company 39 to remain downtown and cover the quarters of Engine 25 and to respond to all alarms downtown. Engine 1 covered Engine 14 and Engine 41 covered Engine 28. Thus, in the event of a fire, at least one company responding would be able to activate these special hydrants.

Currently, the site of the Plant Shoe complex is occupied by a large supermarket that serves local families in the now largely Hispanic community.

[T.G.]

5. Bellflower Street Conflagration

Bellflower Street, Dorchester

On May 22, 1964, the Boston Fire Department was challenged by one of the largest fires in its history when a Dorchester neighborhood erupted into a raging conflagration on a quiet spring afternoon. During 1963 and into 1964, the Northeast section of the United States was experiencing a severe drought. Water bans were placed in effect for nonessential use. Most of Boston's housing stock is wood, with a type of construction unique to the Northeast: three-deckers, or wooden three-story dwellings with wood porches on each floor that extend across the rear of the building. The prolonged drought had made the already fire-prone three-deckers tinder dry.

On the afternoon of May 22, Boston was experiencing a warm day with low humidity and winds blowing in from the southwest with gusts to thirty-one miles per hour. The Bellflower Street section of Dorchester is lined with well-maintained three-deckers. Shortly after 1:30 p.m., trash collectors were making their way through the neighborhood. Suddenly, they spotted smoke coming from the rear of 26 Bellflower Street. They alerted the residents and sounded the alarm. At 1:38 p.m., the switchboard at Boston Fire Alarm Headquarters received the first of many calls from a woman reporting a fire at 26 Bellflower Street; she ended the call saying, "I've got to get out of here!" The switchboard lit up with a surge of calls reporting a serious fire in buildings on Bellflower Street and several fire alarm boxes in the area were received. Engine 21, which was just leaving quarters for in-service inspection, and Ladder 20 were dispatched and Box 7251, at Dorchester Avenue and Dorset Street, was transmitted at 1:39 p.m.

As Ladder 20 responded over Southampton Street, Lieutenant James D. Kennedy, a seasoned Boston smoke eater, could see frightening conditions: a wide plume of black smoke rising rapidly and expanding from the densely packed neighborhood. Even though he was not at the scene, he ordered a second alarm based on what he could see. Responding from Engine 39's quarters on the Congress Street side of District 6 was District Chief John R. Greene. While responding, he could see the ominous black loom up ahead of him. Kennedy arrived on Bellflower Street to find a row of three-deckers heavily involved in fire erupting from the second and third floors of 24 and 26 Bellflower Street. The fire was already threatening the row of three-deckers across

A major conflagration occurred on May 22, 1964, in the Bellflower Street area of South Boston and Dorchester. Nineteen buildings were affected on this warm, windy, dry spring day. Extensive mutual aid from communities outside Boston worked at this fire.

narrow Bellflower Street. He ordered Ladder 20's driver to keep the ladder truck out on Boston Street to make room for Engine 21 and Engine 43 to get into Bellflower Street. Radiant heat was spreading the fire in all directions.

District 6, arriving shortly after Ladder 20, made a rapid size up and Chief Greene, sensing the seriousness of the situation, ordered the third alarm. He began directing companies to cut off the rapid extension of the fire across Bellflower Street to the northeast. Several blocks of three-deckers were in harm's way. As soon as the third alarm stopped striking, Chief Greene ordered the fourth alarm. He ran down the alley between two of the three-deckers on Bellflower Street to better appraise the situation in the rear of the fire buildings and found that the fire was extending in all directions—even against the wind, to the southwest—where three-deckers on Dorset Street were igniting. He attempted to order the fifth alarm, but was unable to break through the heavy radio traffic. The fire had jumped Bellflower Street, and Chief Greene was unable to get back to the Boston Street side of the fire. At 1:45 p.m., Deputy Fire Chief Frederick P. Clauss, Division 1, arrived and ordered the fifth alarm a minute later.

The fire was raging along both sides of Bellflower Street toward Boston Street and Dorchester Avenue, igniting the rear of three-deckers on Dorset Street and threatening homes on Howell Street to the north. Boston had a conflagration in progress.

Repeated calls for additional help were made, with companies directed to respond into Dorset Street and operate heavy-stream appliances in the rear of the buildings. They were also directed into Bellflower Street to operate heavy-stream appliances and big lines in an attempt to halt the raging conflagration.

On the leeward side of the fire, fully involved three-deckers on the northeast side of Bellflower Street were creating a severe exposure problem in the rear of three-deckers on Howell Street.

Acting Chief of Department John E. Clougherty arrived at Boston and Howell Streets and ordered additional engine companies into Howell Street, an assignment that was mostly filled by companies from outside of Boston. Assistant Fire Chief James J. Flanagan arrived two minutes later and took command of the Dorset Street and Dorchester Avenue sector, followed by Chief of Staff John F. Howard taking command at Boston and Dorset Streets.

According to the Boston Fire Department official report on the fire, "The driving wind was whipping the flames from the fronts of the structures on the odd-numbered side of Bellflower through the alleys between the buildings creating a forge-like vortex of fire which viciously twisted upward consuming all material components it contacted and generating extremely high temperatures."

Repeated calls for more fire companies included a dire call at 2:00 p.m. to "send everything available to the fire." At 2:02 p.m., another dire call came: "Send mutual aid to the fire and send all possible help to the fire." Mutual aid—companies from outside of Boston—were sent directly to the fire.

At 2:05 p.m., nine three-deckers on Dorset Street were reported to be fully involved. The acting chief of department determined that classical tactics would be used to attempt to stop the inferno: locate, surround and extinguish. In addition to the exposure fires extending in all directions from the three-deckers on Bellflower Street, flying brands rained down upon buildings at Howell, Rawson and Washburn Streets, as well as Dorchester Avenue. Ladder companies were directed into these areas where, assisted by mutual aid and citizens helping out, they performed yeoman's duty, keeping those fires in check.

Unlike the Great Boston Fire ninety-two years earlier, the water supply system essentially remained sufficient despite the great demand placed upon it. Throughout the fire, engine companies connected to hydrants and relayed water in from all directions, resulting in good hose streams, the essential ingredient in containing the conflagration.

At 2:45 p.m., Clougherty ordered Groups 6 and 7 to return to duty. With sufficient streams concentrated on the Dorset and Howell Street sides, it was determined to initiate a flanking attack by firefighters in the rear yards between Bellflower and Howell Streets at the Dorchester Avenue end. The main body of fire must be held there or it would spread to the next firebreak: the Southeast Expressway. Firefighters channeled and compressed the fire along the northeast and southeast edges. Heavy streams were

Seven newly appointed firefighter candidates use pompier ladders to scale the outside of the drill tower at the Bristol Street headquarters, circa 1930. This exercise demonstrates their confidence in their equipment.

directed on the main fire rather than exposures, since the cooling effect was much greater on the actual fire than on exposed surfaces. Facing punishing heat and choking smoke, firefighters wet themselves with hoses, reversed their helmets and crouched in puddles to hold their positions. After waging a valiant battle, and being driven back several times, they gradually gained the upper hand and checked farther spread of the fire toward the northeast

At 3:30 p.m., nearly two hours into the battle, the first signs appeared that the fire was slowly being contained. While it still raged, the fire was losing strength, but much work remained to complete extinguishment.

The response had been massive: thirty-three Boston engine companies, seven Boston ladder companies, one Boston rescue company, twenty-four mutual aid engine companies, three mutual aid ladder companies and one mutual aid rescue company. The fire flow was estimated at 19,300 gallons per minute.

The fire had damaged twenty-eight three-deckers, seventeen of which were totally destroyed. Starting in the rear of 26 Bellflower Street, the blaze ultimately spread to 16 through 30 Bellflower Street, 17 through 29 Bellflower Street, 21 through 41 Dorset Street, 22 through 26 Howell Street and 140–142 Boston Street. Miraculously, there were no fatalities or serious injuries.

A new Bellflower Court, with housing for the elderly, occupies the primary fire site; otherwise the neighborhood remains much as it was before the fire. The memory of that terrible, warm spring day in 1964 lives on only in the memory of its oldest residents and the firefighters who fought that conflagration.

Firefighters who fought the blaze had this thought: "After this, they'll only be small ones."

[P.C.]

ABOUT THIS BOOK

The Boston Fire Historical Society (BFHS) was founded as a Massachusetts nonprofit corporation in 2006. Currently it consists of a nine-person Board of Directors whose purpose is the preservation and interpretation of the history of the Boston Fire Department. In addition, the BFHS has a Council of Advisors that supports the board on an ongoing basis. Each board member and council member brings to the organization a unique piece of expertise that contributes to the overall mission of the BFHS. At its inception, the organization adopted a five-year strategic plan that is intended to roll forward one year, each year, and is subject to constant review as conditions evolve.

The creation of this volume was an ambitious undertaking, coming as it does in our first year. It has been a total group effort, made possible by the expertise of the several board members and nurtured by their love of the Boston Fire Department and the city that it serves.

All photos are from the collection of Bill Noonan, and all maps are by Karen Bradley.

Sources

Boston Fire Department fire reports.
Ditzel, Paul. *Fire Engines, Fire Fighters.* New Albany, IN: Fire Buff Publishers, 1993.
Noonan, Bill. Essays as part of his ongoing fire history research project.
Personal and anecdotal accounts.
Stapleton, Leo. *Thirty Years On The Line.* Boston: Quinlan Publishing, 1983.
Werner, William. *History of the Boston Fire Department and Boston Fire Alarm System.* Boston: Boston Sparks Association, 1974, 1975.

BOSTON IN FLAMES.

An 1872 etching of the Great Boston Fire dramatically illustrated the extent of the fire. In reality, the fire burned progressively across the downtown area of Boston.

ABOUT THE AUTHORS

PAUL A. CHRISTIAN retired as fire commissioner/chief of department of the Boston Fire Department on February 15, 2006, following a thirty-eight-year career. Appointed to the Boston Fire Department in January 1968, Christian has held all ranks in the organization, and achieved the rank of deputy fire chief earlier in his career than any member in the department's history. He was awarded the department's Roll of Merit and Distinguished Service Award for rescues performed in 1971 and 1972. Christian was appointed chief of department in 2000 and fire commissioner in 2001. During his administration, he reorganized the department and established the Special Operations Division to deal with technical rescue, hazardous material and terrorist response and emergency management procedures and training. Following the terrorist attacks of September 11, 2001, he oversaw the restructuring of protocols and procedures to address terrorism-related duties of the department. Additionally, he directed the department's action plans for the 2004 Democratic National Convention and the Patriots and Red Sox playoff and victory celebrations. Commissioner Christian is a recognized authority on the history of the Boston Fire Department and is a frequent guest lecturer at area colleges and civic groups. He is vice-president of the Boston Fire Historical Society.

THEODORE GERBER is an independent insurance and investment broker living in Somerville. Born and raised in Boston, at an early age he developed an admiration for the skill and bravery of the Boston Fire Department and later, an appreciation of the department's history. He is a member of various fire service support organizations, a past chairman of the Boston Fire Museum and founder and president of the Boston Fire Historical Society. He is also a former U.S. Army captain and saw combat in Vietnam.

MICHAEL W. GERRY, a Braintree resident, is treasurer of the Boston Fire Historical Society and dispatcher for the Massachusetts State Police Troop E. He is a fire historian and a past president of both the Boston Sparks Association and the Boston Fire Museum. He is a veteran of submarine service in the U.S. Naval Reserve and a grandson of a late Quincy Fire Department lieutenant.

About the Authors

WILLIAM T. MURRAY was born and raised in Boston and currently lives in the Hyde Park district. He is the retired director of Fire Safety/Emergency Management for the Boston Public Schools, chief master sergeant, military fire chief, 102nd CES, Otis ANGB, of Cape Cod (retired) and has been affiliated with several organizations that support the Boston Fire Department.

BILL NOONAN has been a firefighter with the Boston Fire Department since 1971, when he was appointed to Engine Company 3 in the South End. He eventually transferred to the Fire Prevention Division as a department photographer. He has authored four photo books: *Flames & Faces: A Photographic Essay of the Boston Fire Department* (2004), *Wooden Sticks and Iron Men* (2000), *Jakes Under Fire* (1997) and *Smoke Showin'* (1984). He is a U.S. Army veteran who served one year in Saigon, South Vietnam. He is married with two adult children.

STEPHANIE SCHOROW is a writer and reporter living in Medford. She is the author of *Boston on Fire: A History of Firefighting in Boston* and *The Cocoanut Grove Fire*. Her newest book, *The Crime of the Century: How the Brink's Robbers Stole Millions and the Hearts of Boston* will be published in February by Commonwealth Editions, and she is currently working on a book on the Boston Harbor Islands for The History Press.

JAMES TEED, a lifelong Boston area resident, was a member of the Boston Fire Department from 1970 to 1999. He was appointed to Engine 12 in 1970 and promoted to lieutenant in 1976. He retired in 1999. He has been doing research on Boston fire history for thirty years.

Many thanks to Boston Fire Historical Society board member PAUL BONANNO for his input and expertise, and to KAREN BRADLEY for creating the maps.

Please visit us at
www.historypress.net